Raku Firing

Advanced Techniques

Raku Firing

Ceramic Arts Handbook Series

Edited by Bill Jones

The American Ceramic Society
600 N. Cleveland Ave., Suite 210
Westerville, Ohio 43082

www.CeramicArtsDaily.org

The American Ceramic Society
600 N. Cleveland Ave., Suite 210
Westerville, OH 43082

13 12 11 10 09 5 4 3 2 1

ISBN: 978-1-57498-301-2

Publisher: Charles Spahr, President, Ceramic Publications Company, a wholly owned subsidiary of The American Ceramic Society

Art Book Program Manager: Bill Jones

Editor: Bill Jones

Graphic Design and Production: Melissa Bury, Bury Design, Westerville, Ohio

Cover Images: "Koshare Bank" by Billy Ray Mangham; (top right) "Urn" by Cory Pinassi; (bottom right) Raku tiles by Tom Radca

Frontispiece: "Relic" by Michael Lancaster

Printed in China

Contents

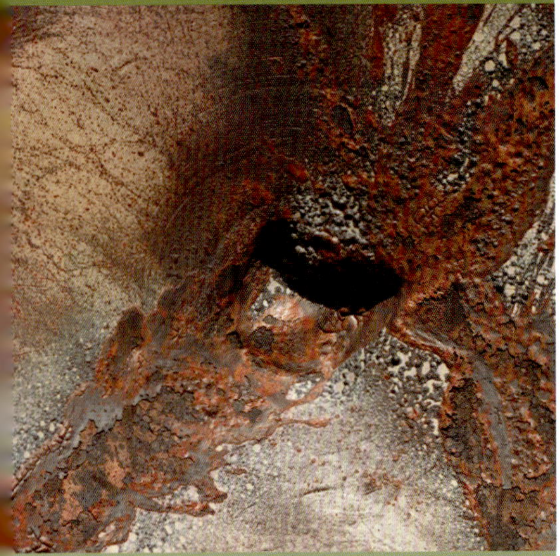

Preface

Raku is probably one of the most popular processes in ceramics due mainly to its simple straightforward process from ball-of-clay to finished glazed ware. And while raku is simple enough for potters of all skill levels to achieve some degree of immediate success, it's the technically skilled and creative artists that have elevated this process to new heights. Raku to them is a vehicle that allows the flexibility to explore and experiment, and in doing so, take the medium to an inspiring level.

This handbook was originally published in the Ceramics Monthly Handbook series in 2004 and soon went out of print. This new revised Ceramic Arts Handbook edition contains more than thirty articles, many of which appeared in the original handbook but with the addition of new materials published in *Ceramics Monthly* and *Pottery Making Illustrated.*

From Hal Riegger, the father of modern day Western raku, to some of the best known raku artists practicing today, you'll find information on forming, finishing, and firing techniques, as well as recipes and stories of the journeys and personal experiences some of these artists have made. For the ceramic artist who has experienced the excitement and the immediacy of the raku process, this book is sure to contain a wealth of information for further exploration and possibilities.

Bill Jones

Raku Then and Now

by Hal Riegger

Raku plate, approximately 9 inches in length, with poured and brushed glaze decoration.

In the spring of 1948, I thought it would be fun to invite a few friends over, have a potluck supper outdoors and try this raku thing. Aside from Bernard Leach's description and drawings of raku kilns and pots in A Potter's Book, the only mention of raku I might have seen was in The American Ceramic Society Bulletin's February 1943 issue. At the beginning of Warren Gilbertson's report on "Making of Raku Ware and Its Value in the Teaching of Beginners' Pottery in America," he mentions (almost casually) that raku's major use in Japan was for the tea ceremony. If either of these sources was where I first heard about raku, I cannot remember. Regardless, I certainly didn't read carefully, or forgot some of what I read.

The weather was perfect for that first raku firing. I lived in the country, partway up Mt. Tamalpais in Marin County, just north of San Francisco. Each of my guests made a couple of tea bowls, and while they were drying, we had a fine supper with good California wine, then got down to the business of finishing the bowls. They were glazed raw, then dried again. I'd made a small electric kiln and put it outdoors. I don't remember if I was first or if one of the guests took the initiative, but a bowl was put into the red-hot kiln and the lid replaced. In about a minute, there was a muffled sort of "poof" sound. We knew what had happened.

I turned off the kiln, and cleaned it out with the vacuum cleaner. A second pot was then put into the

American raku pioneer Hal Riegger removing a plate from the kiln for postfiring reduction.

kiln, which had by then been turned on but was not yet red hot. It took about three minutes before that telltale muffled "poof" again.

"Well, that's it!" I thought. "Those Japanese don't know what they're talking about. The heck with raku."

But I didn't want to give up. Alone, over breakfast the next morning, I reread Gilbertson's article, and there on the very first page, he clearly describes the "biscuit" firing. Maybe the Japanese were okay after all.

More slowly, more carefully, I gave it another try, this time using several little bisqued bowls. Although I had wished for someone to share my excitement, I was alone with my first success: a pinched tea bowl, foot turned on the wheel; the glaze was a pale turquoise with a pink glow inside, and the foot was black.

Before I had an opportunity to do much more raku, a change in jobs took me to Kansas as a designer for a small pottery manufacturer; on return to California, I taught at the College of Arts and Crafts in Oakland. But in 1958, the first time I taught at Haystack, raku was the subject. Those weeks marked the beginning of my serious involvement with the process.

Two years earlier, at the first American Craftsman's Council convention at Asilomar, near Carmel, California, I had met a Japanese potter who had done raku. Of course I quizzed him about all the aspects of raku that I wanted to know. I remember him making a crude little sketch of a raku kiln on a tiny yellow piece of paper. I wish I still had that sketch. Although we didn't follow it exactly, it formed the basis of the kiln we used at Haystack.

During three years of teaching in Florida, from 1958 through 1961, I did some more raku and primitive firings (what people now call "pit" firing). A year after returning from Florida to California, I started what I called Experiment A workshops (the title has absolutely no meaning whatsoever). These were week-long workshops "in the field," where we would find our clay, make and fire pots, all in a very primitive environment. Some of these workshops also dealt with raku fired in wood-burning kilns.

These workshops were held over a period of 17 years in various places around the country, such as the beach at Mendocino, Panamint Valley in the Mojave area, a ghost town in Wyoming and at Cripple Creek in Colorado. Attendees for the primitive pottery workshops were allowed to bring only a shovel and bucket. Other tools had to be found objects.

At first it was hard for us to get totally away from the studio when it came to kilns. Here, we used firebrick and shelves, until one occasion when none were available; it was then that a real breakthrough occurred. An entire wood-burning, top-loading kiln was made from a mixture of clay, sandy gravel and sawdust in more or less equal proportions. We also discovered that a fire could be lit in the wet kiln to dry it out and the mixture wouldn't explode. Although slower than subsequent firings, the first firing took place the day after the clay/gravel/sawdust kiln was built. Some of the kilns the students built became almost human sculptures with breasts and buttocks.

At any rate, by 1964, we had moved a step further from the limiting influence of studio equipment and routine processes into situations demanding creative thought and inventiveness arising from the situation. And, for raku, this was all to the good.

As any potter is wont, when presented with a new tool, material or technique, the workshop participants would go all out and try all possible variations. In retrospect, some of the objects produced may not be quite appropriate in the traditional raku context, but we did explore and we did have fun.

Pinched tea bowl, approximately 4 inches in height.

Raku tea bowl, approximately 4 inches in height, pinched, partially brushed with glaze.

I'm at a loss to explain why or even when I did it myself, but one of the things not traditional among raku potters in Japan that is common among Western potters is a technique called "postfiring reduction." In other words, the pot is put into a reducing atmosphere as it comes out of the glaze firing. This is usually done by placing the pot(s) hot from

A wood-fired raku kiln built from a mixture of fireclay, sandy gravel and sawdust.

the kiln into easily combustible materials in a garbage can and covering with the lid.

Somewhere along the line, I chose a different way that I like better: A small amount of sawdust is sprinkled on the ground to receive the hot raku object. A flat object (a plate perhaps) is just laid on the sawdust, and after a short interval, more sawdust is sprinkled on top of it. If rakuing a hollow form, such as a tea bowl, I wave it in the air a few seconds, place it on the sawdust and push sawdust up to the rim on the outside only. After perhaps a minute, more sawdust is sprinkled on, burying the whole object. This wait is necessary, as the inside glaze cools more slowly and must solidify or sawdust will mar its surface.

Western postreduction techniques may well have arisen from a misunderstanding of the traditional Japanese reduction, which was done after the bisque and before glazing. Usually made of a very rough white clay, the bisque-fired tea bowls were brushed with ocher, then placed in a fireclay box along with charcoal, which was then ignited. The whole process is a bit more detailed than what I have written here, but is well described and illustrated in Herb Sanders' book The World of Japanese Ceramics.

After I quit having the workshops, I began to look at where I was with raku, what I was doing, how integral the process was to the result. In other words, were the things I was doing possible only with the raku process, or could they be made using any of a number of other methods available to potters?

To me, this is the key. It's not a matter of what materials one works with, what temperature one works at, or whether the firing is done with electricity, gas or wood, but whether the object is one that could only be made with the raku process.

If I wanted to go further, following the more traditional ways that tie in with Zen philosophy, I could ask of this work whether it has the qualities that are so important to adherents of the tea ceremony. While most raku potters in America do not assess their pots in the same manner as a Japanese Zen master, there are aspects of a good ceremonial tea bowl that can apply and are indeed

Raku plate, approximately 9 inches in length, sandy red clay, with clear glaze, borax solution and trailed white glaze, by Hal Riegger.

valuable for the Western potter to understand and assimilate.

The traditional raku tea bowl is handmade from a rough, gritty clay, yet when completed, is pleasant to the touch. A summer tea bowl is shallow and wide, while a winter tea bowl is taller and narrower. For its size, it is thick but not heavy. It is asymmetrical; the rim undulates. There is a front and a back. The base rim (foot) must be so shaped that the bowl is easily picked up by the thumb and two fingers. The bottom inside must have a depression called a cha damari, or tea pool, where the last few drops of tea will look like rain collecting in a depressed rock (the potter consciously puts the cha damari there, yet it is supposed to look like it just sort of happened). The bowl must be larger at the lower belly to allow room for the whisk. And, finally, the inside must appear larger than the outside!

Once, when discussing these aspects with a Japanese raku potter, I remarked, "You know, is this rationalizing? I wonder if this isn't all a lot of hogwash."

To which he replied, "Now you're getting into Zen."

Discovering Naked Raku

by Kate and Will Jacobson

"Spirals," 20 inches in height.

As potters, we once wholeheartedly subscribed to the doctrine of utilitarianism, which was the vehicle for financial stability in our production studio. But, over the last several years, we have dramatically altered our view.

The transition began when we realized the production business had somehow transformed us from potters to manufacturers. Somewhere between purchase-order deadlines and workers compensation insurance we forgot why we had become potters. We were marketing to over 300 wholesale accounts and retailing at 25 craft shows a year. We had a computer to keep track of our files—our pottery had become merchandise, numbers and units. What began as a labor of love became just labor.

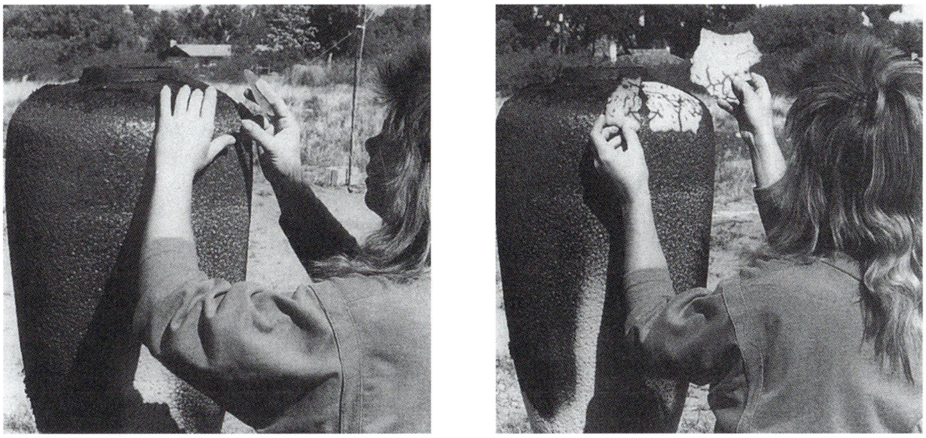

After a Cone 012–010 bisque firing, each pot is covered with Naked Slip, glazed with Naked Glaze, fired to cone 014, smoked; water quenched; and peeled.

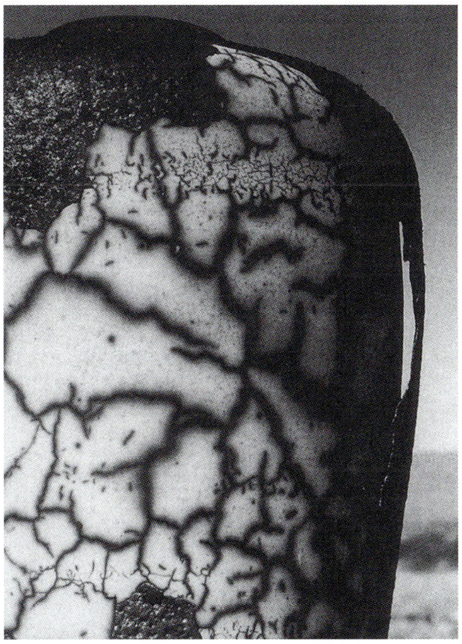

Smoke penetrating the layer of fused slip and glaze creates random carbon patterning that follows crack lines.

As the eggshell-like layer of slip and glaze is loosened, large sheets begin to fall away.

As we improved technically over the years, it seemed our work was losing a sense of self-expression, creativity and joy that drew us to working in clay. Usefulness became an obstacle. Even technical proficiency became a trap. During our neophyte years, clay was the medium for our art. When clay became a business, we lost the fundamental determinant of any art form: the freedom to explore, to experiment, to make mistakes and to pursue the happy accident.

If all we had wanted out of life was financial security, we would have become mid-level functionaries and acquired enough wealth to afford to vote Republican. Something had to

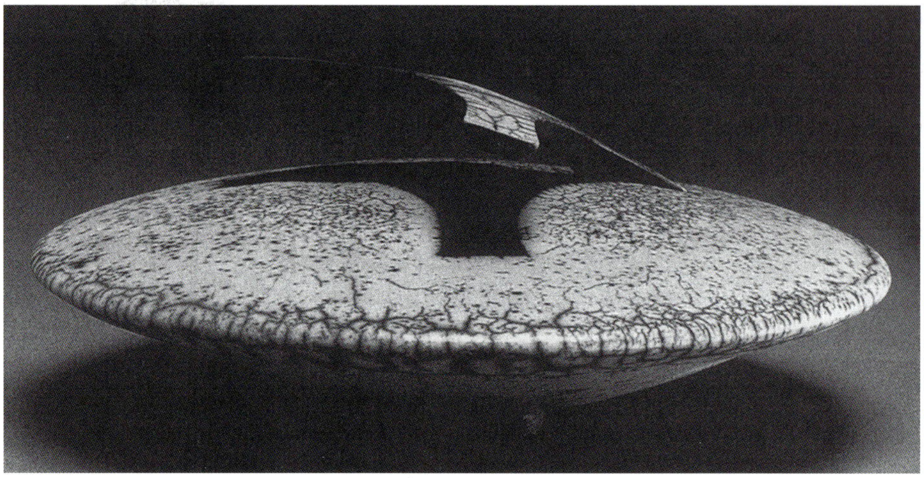

"Gateway," 12 inches in diameter, a closed form thrown from raku clay, with cut and lifted handlelike element, bisqued, covered with slip/glaze, smoke and peeled.

change. We chose to rediscover the art in our craft.

Touring the country with retail craft fairs gave us the opportunity to see diverse contemporary art. In each city, we investigated the local art scene. Museums and galleries were great sources for new ideas, but exchang- ing thoughts with other artists had the greatest impact on broadening our attitudes.

Coming from a functional point of view, we had built mental walls (restrictions) concerning the acceptable parameters of clay. A cup, for instance, could only be made within certain size limitations and remain useful. But there are no pottery police who write citations for illegal use of clay or for parking your acrylic paint on a pot without a permit.

We were determined to create work that reflected our new consciousness, trying anything that came to mind. We willfully wasted time that could have been spent producing salable ceramic goods.

Of course, creativity can be a good deal less lucrative than functionalism and repetition. Even when our explorations ended in artistic success, that did not necessarily mean economic success. For instance, at one point we were experimenting with low-fire salt techniques. Although we and other potters considered them to be phenomenally great pots, we couldn't find a market for them. Also, fully 70% were laid to rest through frustration therapy—throwing them at the dumpster from 20 feet away.

Finally, we found something new and exciting. We were experimenting with colored slips underneath raku glazes, when we noticed a small part of the glaze had flaked off one of our test pots. What was happening under the glaze seemed to be of interest, so we used a knife to scrape all the remaining glaze off the pot, then scrubbed with steel wool.

Rather than dismissing this as a glaze flaw, we considered it a new decorating process. We brush a slip onto bisqued work, let it dry, then apply glaze. As the glaze melts, it fuses to the slip and is prevented from reaching the pot. After traditional raku postfiring reduction, the glaze can be peeled off the pot. The resulting thick shadow lines show where the carbon seeped through the outside glaze layer. We have found we can create a batik effect by layering glaze in repeated firings. Figurative work can be drawn through the glaze, and a reverse image can be achieved with carbon saturation. It took us nearly two years to develop this process sufficiently to reliably control the outcome. The potential applications seem endless.

Maintaining a sense of wonder about our work is absolutely crucial. The freedom to experiment and explore new forms or techniques is a necessary part of our lives. But it is difficult—particularly for potters—to balance creative exploration and economic independence. It is easy to get caught up in the day-to-day routine of repeating successful designs and processes. Our markets perpetuate repetition. Potters share a problem with other artists in that even progressive galleries tend to buy work they've sold in the past and are often wary of new designs.

While artists in other media don't have the initial financial security offered by utilitarianism, neither are they as likely to be seduced (by technical skill and economic success) into just manufacturing. At the same time, it's hard to imagine completely abandoning utilitarianism. There's something very special about a well-made bowl. It feels good to make something both beautiful and useful. There is comfort in the knowledge that we are continuing an ancient and noble tradition. But we also know that clay is a uniquely plastic medium of communication and expression—and there comes a time when technical expertise becomes something to master and transcend.

After all, art and craft, artist and artisan, are just labels that don't have to be kept cleanly compartmentalized. We prefer to approach our craft as artists, with the freedom to explore and fail occasionally and, most importantly, the freedom to discover.

Recipes

Naked Slip

Lincoln 60 Fireclay	40.0 %
No. 6 Tile Clay	30.0
Ione 412 Grog	20.0
Custer Feldspar	10.0
	100.0 %

Apply to bisque.

Naked Glaze

Ferro Frit 3110	60.0 %
Gerstley Borate	40.0
	100.0 %

Apply over Naked Slip.

Marcia Selsor
Raku-fired Slabs

by Louana Lackey

"Mustangs of the Pryors," 40 inches in length, raku-fired diptych.

Marcia Selsor paints large raku plaques with cattle or horses. She started making the slabs after she saw "Black Angus against fresh green grass" while driving home one bright spring day in Montana. "I immediately thought 'raku!' Somewhat later, we took two carloads of students up to the Pryor Mountains to dig for clay. As we drove across the top in a cloud, a herd of mustangs came from behind us, surrounded the cars and ran down into a coulee. That was when I started the horse series."

Process

Irregularly shaped slabs for the plaques are prepared on a slab roller. Selsor then adds four nubbins to the backs of each—two to be pierced for the hanging wire, two to keep the slab an even distance from the wall. After the plaques have been dried and bisqued, she sketches the horse designs using a 6B or 8B drawing pencil (figure 1). Latex resist is then brushed over the whole design (figure 2). She has found two thin coats of latex to be easier to remove than one thick coat. Finally, a light coat of a copper matt glaze called Dry Alligator is sprayed over the entire surface (figure 3).

After about an hour, Selsor uses a needle tool to lift off the latex covering the designs (figure 4). She then redraws the lines with a slip trailer (an ear syringe tipped with a hypodermic needle) filled with a red/bronze copper luster glaze (strained before use) (figure 5). She first used this technique to outline designs in 1995, when she "visualized horses, seen from above, shimmering like ghosts in the night."

Selsor waits a day, sometimes two, before firing. She arranges the pieces on their sides in the raku kiln (figure 6) and fires them to cone 05. On removing them from the kiln, she waits a few seconds before inverting them onto beds of straw and covering them with galvanized tubs (figure 7). A very short wait gives the background brighter color (figure 8).

"Seekers," 24 inches in height, with latex-resisted and trailed raku glazes, smoked in straw.

Sketching designs, using a 6B or 8B drawing pencil, on bisqued slabs.

Applying liquid latex to the pencil drawing as a glaze resist.

Recipes

Red/Bronze Raku Glaze

Colemanite	50.0 %
Ferro Frit 3134	50.0
	100.0 %
Add: Tin Oxide	5.1 %
Copper Carbonate	5.6 %

Dry Alligator Raku Glaze

Bone Ash	20 grams
Colemanite	40
Nepheline Syenite	10
Copper Carbonate	10
	80 grams

Spraying Dry Alligator Raku Glaze over the entire slab surface.

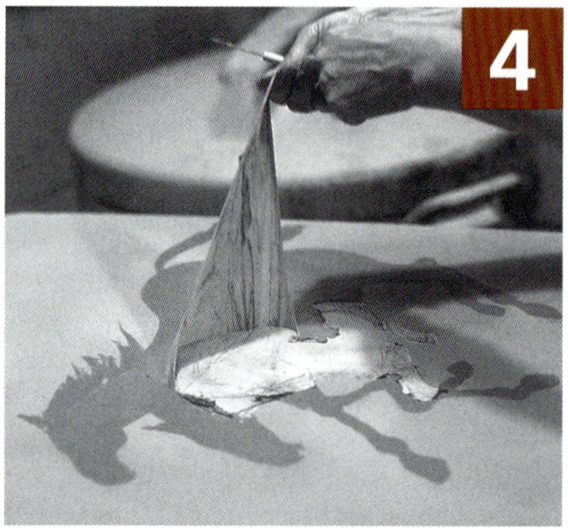

Lifting the latex resist from the dry glazed surface.

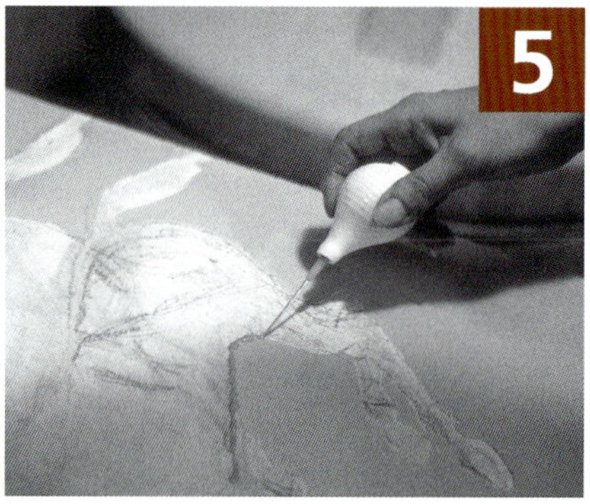

Defining the drawing lines with trailed Red/Bronze Raku Glaze.

Glazed slabs loaded on edge in the raku kiln.

After firing to cone 05, the slabs are removed from the kiln, laid face down on a bed of straw, then covered with a galvanized washtub.

A short smoking period yields carbon-impregnated surfaces wherever the liquid latex had been used to resist the sprayed glaze.

Raku at Amatlan

by Hilda San Vicente Tello

Vase, 21 inches in height, wheel thrown, brushed with raku glazes.

About 50 miles south of Mexico City lies the village of Amatlan de Quetzacoatl. It is situated at the edge of high rocky mountains overlooking lush fields in the valley below. Quetzacoatl is the plumed serpent or feathered snake of the Nahautl people; the name means god of goodness. It was in the shade of Amatlan de Quetzacoatl's 16th-century Augustine chapel that Needham, Massachusetts, potter Steven Branfman conducted a raku seminar sponsored by El Olvido Ceramica.

Located in the heart of Mexico City, El Olvido Ceramica is a high-temperature ceramics studio where a group of 35 ceramists make their best efforts to maintain an educated dialogue with an excellent yet often difficult interlocutor—clay. Several years ago we decided to study the raku technique. With the help of some basic literature, most notably Branfman's book *Raku: A Practical Approach,* a homebuilt kiln for raku and lots of enthusiasm, a series of simple forms were glazed and the first firing conducted.

After a long series of trials to find reliable glazes, the correct fusion temperatures, the effect of kiln atmosphere and the huge possibilities of "smoking," we started down the path of studying texture, shape, color and contrast through experimentation. Now, with our glazing knowledge and understanding more mature, we are convinced that in raku "the skin of the piece rules."

Since the beginning, the help of Steven Branfman has been priceless. His experience of over 25 years follows the tradition of Hal Riegger and Paul Soldner, who popularized raku practice in the United States in the 1960s. After having used Branfman's book extensively, we wanted a personal interchange of ideas and invited him to Mexico.

For Branfman, "raku is a process that offers the best of all worlds. The technique is deeply rooted in tradition. And while its origins serve as a constant reminder to me of where the craft has evolved from, its contemporary incarnation is very differ-

Steven Branfman removing a piece from 1 of the 16 raku firings in a portable kiln at a workshop in Amatlan de Quetzacoatl, Mexico.

ent. So, I can work simultaneously in a traditional technique where all the rules have been established, and a contemporary technique where the rules are constantly in question."

The seminar began at El Olvido in Mexico City where, after an evening slide presentation, Branfman spent two full days demonstrating forming techniques. The participants were able to observe an experienced potter wedging, throwing and carefully preparing surfaces for later glazing and final enrichment through raku firing methods. Sessions were full of intelligent, probing questions and equally thought-provoking answers.

The pots were made from a body composed of Mexican clays (47.6% Hidalgo stoneware, 4.8% Oaxaca red clay and 47.6% Zacatecas fireclay); it produced a white, refractory body with very plastic qualities. Glazing was done with some of Branfman's recipes, which allowed the participants to sample his personal glaze palette.

After a weekend of relaxing and sight-seeing, the group arrived in Amatlan de Quetzacoatl. Many of the wares had been previously glazed, and the plan was to complete the glazing at the firing site with Branfman's instruction and guidance. The atmosphere and the beauty of this place helped produce a working euphoria among the participants that only intensified in strength, enthusiasm and creative activity as the hours passed.

Over the course of two days, using two kilns, we completed 16 firing cycles in each kiln. There was much successful experimentation with many of the possibilities that the raku technique offers.

Testing Variables

The main variables tested during our firings were as follows:

- *Oxidizing firing*—maintaining an ample supply of air to produce an oxygen-rich atmosphere. The air supplied to the combustion is more than what is necessary for complete combustion of the fuel.
- *Reduction firing*—restricting the intake of air to observe the effect of carbon-filled atmospheres in producing an "oxblood" effect with copper glazes. The amount of air allowed to enter is less than the necessary amount for efficient combustion.
- *Salt firing*—introducing sodium chloride through the burner after the fusion of the glaze to produce alterations in color development and surface texture.
- *Postfiring reduction*—removing the pots from the kiln and immediately placing them inside a container filled with combustible material. This procedure is used mainly to produce intense smoking, lusters and crackle glaze effects. We also allowed the pieces removed from the kiln to cool in the air for 15 to 60 seconds, then subjected them to the reduction atmosphere of the container. This procedure permits a partial reduction and produces blue or greenish tones mixed with metallics. A third variation involved removing the pieces from the kiln and immediately subjecting them to intense reduction, then opening the container, exposing the ware to the air to produce copper halo and matt effects.

At the end of the day, red and orange hues filled the sky over the valley as the kilns cooled, and Branfman offered his comments and suggestions regarding the work that had been completed. His final words to the group echoed how we viewed the intimacy and personal nature of his presentation: "An extraordinary collection of ware from a special group of people. You have embraced me and my family with your warmth and friendliness. Our experience here will remain forever in our hearts."

The sum of these emotions was captured in each of the pieces produced there. With the help of clay and fire, a strong relationship between the pieces and their creators was demonstrated. The seminar was an unforgettable experience for everyone, and it was made evident that our pieces reveal, to a great extent, our own hearts.

Inlaid Glass Technique

by Steven Branfman

Left: Bottle, 19 inches high, Basic White Crackle glaze.

Right: Vase, 18 inches high, Rogers White (clear) glaze.

Of the questions asked of me at workshops and demonstrations, the most frequent center on the subjects of personal style and the generation of ideas. How to develop a distinctive appearance in your work and how to continue to come up with new ideas can be an intimidating and even forbidding responsibility. Making a personal statement in clay does not demand special attention.

Regardless of what your pots look like, almost from the start, your own identity and ownership of that identity will be apparent. However, keeping your work fresh and your excitement about clay alive may require more deliberate effort. Here are a few conventions to facilitate the generation and discovery of new ideas.

Think less about what your pieces look like and think more about how they feel to you, physically and emotionally.

Set up some technical, aesthetic, or intellectual limitations and boundaries and force yourself to work within them.

Simply try out some method—forming, surface treatment, glazing, or firing—that you have never done or had much success with before. See it through to the completion of a minimum of, say, 10 objects, and see where that exercise takes you.

It also helps to study, understand, and come to grips with the whole theme of ideas, originality, expression, and personality. It has been said, and I agree, that there are no new ideas, only new combinations of things that have already been done.

PROCESS PHOTOS: JARED BRANFMAN

Combining Clay and Glass

I began incorporating glass into my work years ago as an effort to unite two apparently similar materials. Clay and glass are at least cousins if they are not siblings, and although clay was the material that I was connected to, glass held a certain interest.

My first efforts involved incorporating the glass as "windows" through the surfaces of my pots, connecting the glass in the fashion of leaded-glass technique with clay cutouts in the wall of the vessel acting as the frames for the cut pieces of flat glass. At the same time, I was using bits of glass as decorative elements by laying them flat inside bowls, dishes, plates, and platters, and allowing them to melt into the surrounding glaze. Combining these two materials proved to be much more difficult than I had thought it would be. These experiments with glass took place along side other work that was proving to be much more interesting and engaging and soon my interest in the glass all but disappeared.

It wasn't until some years later that the glass idea re-emerged, although from a completely different perspective. From very early, my pots have been about the relationship between the surface of the ware and the form. As I matured as a potter and a maker of vessels, my understanding, interpretation, and representation of the vessel became, and continues to become, more precise, sophisticated, and personal.

How the interior space of a pot shapes, defines, and gives life to what we see on the outside is elemental to how I conceive and make my ware. I form pots from the inside out and the bottom up. And while pressure and force on the outside of the clay during the throwing process is a key element to controlling the shapes and sizes of the forms, it is the inside pressure that I exert that actually creates the pot.

The surface of a pot is more than a simple canvas upon which to deco-

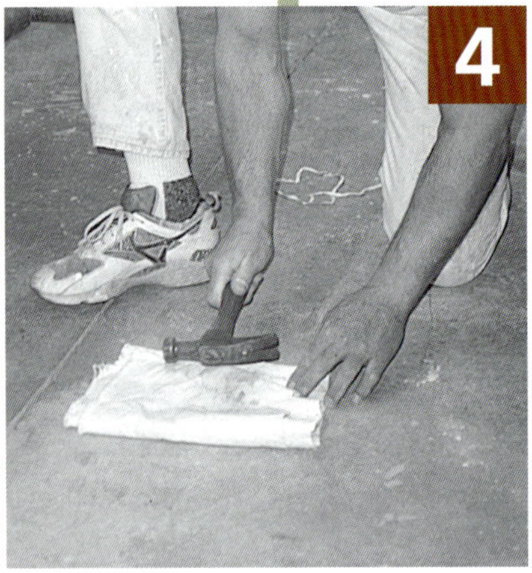

rate or embellish. It is a skin that contains and thus expresses and communicates all of the power within. The incorporation of crushed and inlaid colored glass is yet another method I use to articulate and embody that power.

As with most craft methods, there are many nuances of technique, method, and personal style that are not possible to demonstrate within the limitations of the printed word. Start your experiments with small amounts of clay and simple shapes building up to larger, more complex forms as your skill improves. Don't be satisfied until you've incorporated this new method into your own language of shape, color, texture and form.

Process

There are four steps to the technique: Forming a cylinder, inlaying or embedding the glass, forming the shape by expanding it from the inside out, and finishing the piece.

Although my work is all thrown, with modifications you can easily adapt this technique to slab work.

You must throw your form on a bat (figure 1). Plastic bats that tend to bend are not recommended until you have more experience with this technique. (You'll see why later.) After using liberal amounts of water for centering and the first few pulls, I dry my hands to throw the rest of the clay completely dry.

Note how little water is visible in the photos (figure 2). Dry throwing leaves the cylinder strong and able to withstand the stresses that the clay will have to with stand later in the process. The cylinder must also be formed with very even walls that are left thick enough for the glass to be embedded and subsequently expanded. Dry throwing is a skill all to itself and one that will take a lot of practice to master.

When you complete the forming, the clay should feel somewhat stiff (though not nearly leather hard) and dry to the touch (figure 3). Depending on how dry you throw, you can put the cylinder aside until the surface wetness dries. If you do that, be sure

to maintain even drying, not allowing the cylinder to warp or bend and not allowing the rim to get dry. Tip: Before you lift the bat off the wheel, mark the pins and corresponding holes in the bat so you can put the bat back on the wheel in the same orientation.

While the cylinder is setting up, I prepare my glass. I use random varieties of flat, colored glass used by stained glass workers. You can use any type of glass available to you, including bottle glass, marbles, and glass rods. Wrap the glass in canvas and simply crush it with a hammer (figure 4).

On the table I lay out my glass carefully choosing colors, sizes of glass (from large pieces to dust), and the arrangement of the glass in patterns and shapes visualizing how these patterns will transfer to the surface of the clay (figure 5).

The cylinder will be rolled onto the glass, so as you lay it out pay attention to how close to the rim and the foot the glass will be inlaid (figure 6). You must be able to finish throwing the neck and rim of the piece without touching the glass and then trim the bottom and foot (if you choose to trim) without running the tool over the glass.

Holding the bat almost as a steering wheel and supporting the cylinder from the inside, roll the cylinder in the glass (figure 7). This is where a rigid bat and stiff clay is critical.

Place the cylinder upright and paddle the glass into the clay still supporting the clay from the inside to maintain the integrity of the cylinder (figure 8).

Each time the clay is rolled onto the glass, it's to enhance the pattern and the coverage on the vessel's surface. Repeat this process until the surface is as you want it (figure 9).

The more you roll the cylinder, the more it will become stretched and distorted. Work carefully to minimize the distortion (figure 10).

To combat distortion, gently return the cylinder to its vertical form by grasping the cylinder around the outside, pushing it and pulling it back into shape (figure 11).

In preparation for continuing the throwing process, collar in the top of the cylinder (figure 12). Throw the top of the cylinder slightly to re-orient and re-center it (figure 13). CAUTION: Before you continue, clean your wheel and workspace of all slip, trimmings, and other clay that you will want to recycle. Pieces of glass commonly fall off the form during the expanding step. Return the bat to the wheel and align the pinholes to their original orientation.

If you've never done any one-handed throwing, this next step will take some practice to arrive at a comfortable personal approach. The more centered (meaning evenly thrown walls) the cylinder is, the more success you will have in one-handed throwing (figure 14). I use my left hand inside the form to raise the clay and push it into the shape that I want (figure 15). Yes, even with only one hand you can still dig into the clay and raise it! Hold and control your left hand and fingers just as you would if you had your right hand also working on the outside. A slow, careful approach works best. I alternate between using my hand and using a rib to expand the shape. Although I use my left hand inside the pot for throwing, I am more comfortable and have more control using the rib in my right hand.

When you use a rib, be careful to hold it as flat as possible (not at a 90-degree angle) against the inside wall so you don't scrape clay away (figure 16). The idea is to push the clay outward, not to scrape clay away and unnecessarily thinning the wall. As you expand the wall and it gets thinner, you'll feel the texture of the glass. It will take practice and experience to learn how much you can expand the vessel before you stretch it too much!

With the body of the piece finished, you must move to the neck and rim (figure 17). Throw these sections in the conventional way with left hand in the inside, right hand on the outside. I continue to throw dry, but at this point you can use water if you are careful to restrict the water to only the surface of the clay that you are shaping.

13

14

15

16

All of my work has a trimmed foot ring that I complete when the clay is leather hard (figure 18). Finish your piece according to your own style. Here, my bottle form is inverted in a chuck made immediately before use. The chuck is shaped to fit the form. While you want the chuck to stick slightly to the pot, you don't want it to stick too much (again, only practice and experience will tell), so before placing the vessel in the chuck, dry the rim of the chuck lightly with a towel.

The surface of the finished vessel reveals how expanding the form from the inside out imparts a unique character to the surface of the vessel (figure 19). The areas around the larger pieces of glass have stretched and redefined themselves and the placement of the glass in the clay moves into dynamic patterns.

Glazing and Firing

There are no restrictions on glazing methods and application can be by pouring, dipping, brushing, or spraying. Most of my glazes are brushed on with inexpensive wood-handled Chinese bristle paint brushes (2 to 3 inches wide), available at any home

WARNING

Keep that outside hand off the glass!

Recipes

Basic White Crackle

Cone 08

Gerstley Borate	65 %
Tennessee Ball Clay	5
Nepheline Syenite	15
Silica	5
Tin Oxide	10
	100 %

Rogers White (clear)

Cone 08

Gerstley Borate	60 %
Spodumene	35
Tennessee Ball Clay	5
	100 %

Del Favero Luster

Cone 08

Gerstley Borate	80 %
Cornwall Stone	20
	100 %
Add: Copper Carbonate	2 %

Yellow

Cone 08

Gerstley Borate	80 %
Cornwall Stone	20
	100 %
Add: Vanadium Stain	6 %

The depth of yellow will vary depending on the stain that you use. Add up to 5% tin oxide for a more opaque quality.

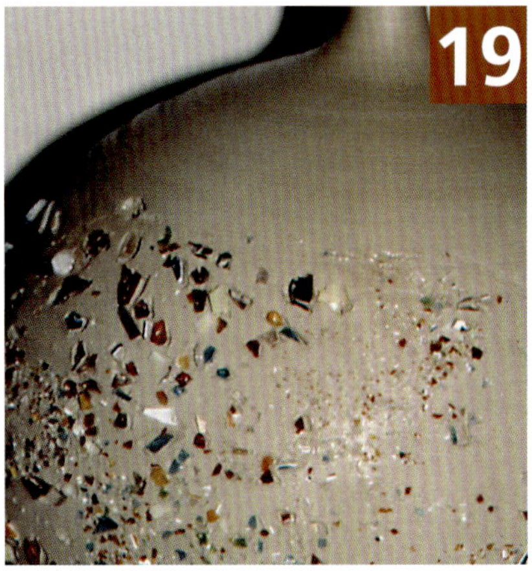

center or paint store. The bristles are soft, they hold a reasonable amount of glaze, and it is comfortable to use as a tool.

Colors, patterns, and textures created by the glass covered with a clear or white glaze are design elements strong enough to stand on their own. However, depending on the effect I want to achieve, I use glaze of different thicknesses to vary the melting of the glass and glazes of different varieties to add additional color and texture.

Glass by itself melts only minimally at bisque temperatures but melts more the higher you fire. In addition, glaze acts as a flux, melting the glass further. Imbedding glass is not a technique tied to raku. If raku doesn't interest you, experiment with different firing styles, glazes, and firing temperatures.

Postfiring reduction is done in metal cans with coarse sawdust, wood shavings or dry pine needles.

Bottle, 20 inches high. Imbedded glass technique glazed with Rogers White with brushes of Del Favero Luster. Slight postfiring reduction.

Successful Raku

by John Ramer Sherrill

"Spring Swallowtails," 15 inches in height, with clear glaze over polychrome slips, and burnished lid carbon-blackened during post-firing reduction, by John Ramer Sherrill.

Over the past couple of decades, raku pottery has become tremendously popular in the United States. The wide range of glazing and firing methods, and the surprises that lurk in every firing hold the interest of potters year after year.

Many achieve consistent results, but I have talked to dozens of established potters as well as students who have been unhappy with their raku attempts. Most complained that they had been unable to find specific information to properly guide them in their efforts.

For my own early efforts, I obtained several books on the subject, but found them frustratingly long on philosophy and short on technique. I still couldn't properly fire a raku pot, but I could use my new-found knowledge of Zen to cope with the situation. I don't believe there are many raku enthusiasts who are interested in my philosophy, wondrous as it may be, but I know for certain that some want to know how to do raku.

Clay Selection

A wide variety of clays can be used to make raku pottery. Be aware, though, that the clay used determines much of the character of the finished piece.

As a general rule, earthenware clays should be avoided if you like dark, bold crackle effects. The closer a clay body gets to vitrification, the less effect carbon will have on it in reduction. Iron content also resists carbonization, so terra-cotta clays should be avoided.

We are then left with the light-colored stoneware clays, and nearly all of these will work well. A stoneware body with just enough iron to give the fired piece a light tan color will have a pleasingly warm appearance. Some of the white-firing clays have an attractive ivory appearance. Experiment with different bodies to find one you like.

Most suppliers sell a body designated as "raku" clay. This is usually a grogged clay that includes Kyanite.

It is the clay of choice for really massive raku pieces; however, I never use "raku clay," nor do I recommend it for general use for the following reasons: 1) the grog content can leave your hands bleeding after long wheel sessions; 2) the grog content breaks surface tension in the glazed piece and causes small "cross-hatch" crackle patterns that I find boring; 3) the body color is usually anemic; 4) besides, with ordinary-sized pieces, most stonewares survive the firing as easily as raku clay.

Suppliers also usually have bodies that are designated as "ovenware" clay. These clays, which contain less grog (easier on the hands), often make an ideal raku body. I use ovenware clay almost exclusively.

Forming and Drying

Raku pots may be wheel thrown, handbuilt or cast. I have heard from several sources that cast pieces cannot be raku fired, but they are no problem.

Some consideration needs to be given to proper drying. As a general rule, drying pots of 3 pounds or less does not require special handling. I often force dry and bisque fire the same day. Flat pieces and large pots must be dried slowly and evenly, though. Large ovenware pots will often survive fast drying, but the stresses that are thereby induced will cause them to crack at a later stage.

Slip Decoration

If you want a colorful pot, you may use oxides or stains in the glaze, but they may mask the dark crackles to some degree; some almost entirely obliterate them. For that reason, I use colored slips under a clear glaze. My slip base is simple—1 part ball clay to 1 part EPK kaolin. Just mix it with water to a cream consistency and add stain. I use commercial stains in percentages ranging from 2% (dark blue and green stains) to as much as 30% (pink stains).

In order to choose the stains that will work well, it is good to know their chemical components. Most commercial stains will block carbon to some extent, with the worst offenders being those that contain iron or vanadium. Vanadium is present in most warm-tone commercial stains, so you should use titanium yellow, praseodymium yellow or zirconium yellow in combination with other vanadium-free stains to formulate your own palette.

Ideally, the slip should be brushed or dipped on at the leather-hard stage, but it works on bisqueware as well. When brushing, you should apply three coats for dark colors (when you don't want the body showing through) and two coats for lighter colors (a warm-tone body showing through some, such as pale green, can be very attractive). The pot should then be bisque fired in the cone 08 to cone 04 range.

Glazing

After bisquing, the pot is ready for glazing and the final firing. Glaze should be applied fairly thickly. If you dip, the consistency should be about that of thick cream, and one dipping should suffice. If you brush, the glaze should be somewhat thicker, and two or three coats should be applied. Evenness of application is not particularly important.

It is somewhat traditional to leave the area near the base unglazed. This will turn quite black in a good postfiring reduction, when carbon penetrates the still-hot pot. Other areas may be left unglazed as well. These areas may be random or symmetrical, and can greatly enhance the beauty of the finished piece.

Kiln Selection

Because red-hot pots are removed from the kiln, it is apparent that many models simply are not appropriate for raku firing. Large kilns of any type, when opened at temperature, radiate heat so fiercely that it would be foolhardy to attempt rakuing. Top-loading kilns are also impractical, as you must position yourself above the kiln in order to reach inside, and the rising heat would be overpowering.

Small (2 cubic feet or less) front-loading electric kilns may be used, but most raku firing is done in gas kilns especially constructed for that purpose. They can be purchased commercially, or they are easily homemade.

Firing

Raku kilns, unlike conventional kilns, are usually loaded on a single level, and spaces between pots are left a bit wider in order to facilitate their removal. It is certainly possible to use multiple levels, but it isn't worth the hassle, as far as I am concerned.

"Webbed Jar," 14 inches high, wheel-thrown ovenware clay, raku fired; blackened areas were taped to resist the glaze.

Traditionally, the maturity of the glaze is determined visually. The kiln is opened, and the pot surfaces are examined for complete glaze melt. If mature, the glazes will appear wet and reflective. I strongly recommend using a pyrometer in conjunction with this technique, noting the temperature at which maturation occurs. After a few firings, you will need to look only at the pyrometer to determine unloading time. There is some medical evidence that prolonged or repeated staring into a red-hot kiln can damage vision.

When examining the pot for complete glaze melt, look for bubbles in the glaze, as these can mar an otherwise perfect pot. Even if the bubbles burst when the kiln is opened, unsightly craters will remain. Bubbles are almost always present on my pots because I fire rapidly, so I simply assume their presence, and take steps to remove them.

To accomplish this, partially open the kiln just long enough to drop the temperature by 200°F or so (I give it about a 10-second count). Close the kiln and bring the temperature back up to near maturity. Give it a couple of minutes for the craters to heal. If you have clusters of bubbles, you may have to repeat the procedure.

Postfiring Reduction

The final phase of raku firing requires the still-hot pot to be placed in combustible materials inside a fireproof receptacle that can be covered, the tighter the better. The combustible material can be sawdust, straw, leaves, newspapers or anything else that readily catches fire. I prefer a bed of sawdust covered with crumpled newspapers, but I suggest trying different materials to discover what best suits you. In any case, the bed of combustibles should be prepared in advance of the firing.

Postfiring reduction is where the novice usually runs into problems. It is potentially a dangerous process, so always take precautions and exercise extreme care. You will be working closely with temperatures up to 1800°F, so you must train yourself to touch nothing without first considering whether or not it may be hot. After a long rakuing session, I actually catch myself hesitating before entering my home, considering whether or not the doorknob is hot. It is a useful habit to cultivate.

Cover as much of your body as possible (always wear long sleeves), but don't wear polyester. Taste in clothing is not the problem; the problem is polyester will melt and conform to your body like hot glue, heat-resistant gloves are a must. A hat and face mask are not absolutely necessary, but are a good idea. I prefer to remove large pots by hand, but for this, special heat-resistant insulated gloves must be used. For smaller pots, long metal tongs are suitable.

Reduction techniques vary quite a lot, so I will simply describe my own; modify as you wish. I remove the pot and place it on some fireproof surface, then wait for cracks to appear in the glaze on the rapidly cooling surface. In bright sunshine, these cracks appear as shiny lines. They will announce their appearance with audible pings or pops.

Only then do I place the pot into the reduction receptacle. This action takes place for a small pot, such as a bud vase, in as little as 10 seconds. A very large vessel (5 pounds and up) may require 90 seconds or even longer. This timing from kiln to postfiring reduction is very important, as it will, in large part, determine what kind of crackle effect will be achieved. I find that the sequence I have described gives deep, dark, widely-spaced crackles usually interspersed with networks of finer lines. Varying the timing should soon show you how to get the effect you prefer.

Quenching

A lot of pots are lost in the next step, which is called "quenching." After reducing for a minute or more, the pot is removed with tongs and submerged immediately in a container of water. The water hisses and bubbles, and the hot pot rolls about as if in pain. Those with narrow mouths will gyrate wildly, and will sometimes rocket themselves clear out of the container. All this commotion by a suddenly animated pot is undeniably a lot of fun, but I no longer enjoy it because I no longer do it.

These days, I just leave small pots in the tightly covered reduction chamber for about 5 minutes, remove them and place them on the ground to cool. I leave large pots in reduction for up to 30 minutes, long enough for them to drop below the quartz-inversion temperature (1063°F), because that is when a large exposed pot is in serious danger of cracking. As far as I've been able to determine, the only thing I lose by not quenching pots are pots and, yes, a bit of fun.

After the pot cools enough to handle, all that remains is to scrub it vigorously with a metal pad or wire brush.

Recipes

The following gloss glazes are dependable and work well with underglazes and colored slips.

Clear Raku Glaze

Cone 06

Gerstley Borate	70 %
Plastic Vitrox Clay	30
	100 %

Translucent White Raku Glaze

Cone 08

Gerstley Borate	80 %
Cornwall Stone	20
	100 %

Clear to Cloudy Raku Glaze

Cone 08

Gerstley Borate	75 %
Feldspar (Soda or Potash)	25
	100 %

Billy Ray Mangham

Reconnecting with the Species

by Gary C. Hatcher

"Koshare Bank," 20 inches in height, wheel thrown and handbuilt.

Raku, wood firing and salt glazing are similar in that the process can obscure content, making process the only element existent in the work. These methods of process domination only work if there is content in the form, which needs enhancement by the process. Bringing form, content and process into balance is the dance all clay artists struggle to perform.

Content transcends process in the raku works of Texas artist Billy Ray Mangham. Mastery of the firing and glazing processes is evident, but all elements are considered and all are brought into harmony. He may make it all look easy, but easy living has been far from the truth for Mangham. Underneath the amiable persona is a highly evolved ethos that provides humor as well as power to his work.

A 30-year pursuit of communicating deep metaphysical and societal truths via art has been his quest. The playful first appearance of the work is the hook. After the viewer is caught, the work's true essence begins to be revealed. This duality of humor and pathos is seen clearly in his "fool" series, comprised of clownlike busts that are strikingly comical. Upon closer inspection, the viewer sees short haiku-like verse on the backs, which signal the "fool" is caught in a web of his own making. Mangham interprets the "fool" as one who stands outside of reality and time, and comments on events witnessed. The viewer wades into the shallow water enticed by humor, then realizes that this work is far deeper.

The son of a hard-working oil company man, Mangham was born on a hot July day in 1945 in the small west Texas town of Phillips. He grew up living in an oil-company house near eight other families. According to Mangham, "All of the families were fairly uneducated oil field 'rednecks' except for the Olds family."

Willeta and Gene Olds made a difference in Billy Ray's life. Willeta taught art at the junior high school and Gene was a rock hound, tinkerer and maker of things. It was at their house, which was filled with exciting junk, and in Willeta's ninth-grade art classes that he found a toehold for a life in art. Their generosity in

"Big Wheel Raku with Found Objects," 28 inches in height, wheel thrown and handbuilt.

"Poultry Cult Figure," 42 inches in height, wheel thrown and handbuilt.

helping him work with tools and his imagination to make wooden toys provided the core for the playful raku sculptures of today.

The seeds of art were sown by Willeta and Gene Olds, but it would be many years before Mangham would be able to orchestrate his life and direct his energy to the realization of his own creative voice. In 1964, he started playing rock and roll. "I played rhythm and blues in honky-tonks in Amarillo during the week, and teen dances all over the panhandle, Oklahoma and New Mexico on weekends. I was supposed to be going to college but ended up flunking out."

Then came a four-year tour of Japan, Hong Kong, Thailand, Hawaii and the Philippines, courtesy of the U.S. Navy. "Vietnam was hot in '67

PHOTOS: ERIC BEGGS, LORI LYNN

Billy Ray Mangham transferring a glowing hot truck sculpture to a trash can for postfiring reduction.

Much of his work begins as a thrown cylinder, which becomes the body for cubist figures, school buses and trucks.

and '68 and I was in serious danger of the draft, so I joined the Navy."

After completing his travels with the military, he met John Daniel, a sculpture professor at Stephen F. Austin State University in Nacogdoches, Texas. Daniel guided Mangham back to the art track that Willeta and Gene Olds had started him on. Throwing himself completely into creating sculpture, he experimented with every material available: wood, welded steel and plaster, but it was when he touched clay that his life took on a true direction.

At the time, the university had no clay department, so his pursuit of clay led him to San Jose State University in California, where Herbert Sanders and James Olevera were his professors. This is where Mangham was first introduced to raku. "Once a year we had a great 'raku-fest' on a beach north of Santa Cruz. Hundreds of pottery students from local colleges gathered for four or five days, camping and firing in a giant party. It was a raku experience in the truest sense of the word."

After completing a B.F.A. at San Jose State University, Mangham and two friends, Ralph Fulton and Ruth Richards, set up a pottery shop back in Nacogdoches. The quick firing process of raku gave way to pursuit of the Leach-Hamada path. "We made high-fire pots for the masses. We built our own house, tried to grow our own food and made functional pots using ash glazes. Our cry was 'back to the land.' We worked very hard trying to make the dream come

"Gator Krewe with Mascot," 38 inches in height, wheel thrown and handbuilt, by Billy Ray Mangham.

true. We had little money, business acumen or skill, but we had a dream and were not afraid to work. In the end, the partnership was dissolved and I moved on to Austin."

Having completed an M.F.A. at Stephen F. Austin, Mangham was offered the position of potter-in-residence at the Laguna Gloria Art Museum. It was at this time the Leach-Hamada approach to working began to lose his interest. After ten years of looking at earth tones, he began to crave color. The transition from earthy colors to bright raku took several years.

Most of his works begin with a thrown bullet shape, which becomes the body for cubist figures, school buses and trucks. The parts are frequently assembled with ungalvanized finishing nails inserted into clay seams at the leather-hard stage. The clay body is one that has almost no shrinkage, and cracking is rarely a problem.

Each piece is bisqued to cone 08, glazed and raku fired to cone 05. This is close to the temperature limit that these works can be fired without the nails melting. The primary glaze is a 50:50 mixture of Gerstley borate and nepheline syenite; it's combined with stains or commercial low-fire glazes to produce various colors.

When asked what the core of his work is, Mangham responded, "I am continually seeking answers to why I have chosen this path. I am working with two assumptions now. First, clay enables me to focus, concentrate and pay attention. Paying attention and being in the present make me truly alive.

"The second reason I have chosen this path with clay is that it connects me to the memory and history of our species. As a culture, we seem to have lost values and direction important to our survival. When I touch clay, I reconnect with the memory and history of our species."

CJ Buckner
Raku Menagerie

by Henry Atwater

The pieces are fired on a metal rack, which aids in removal from the kiln.

The rack and pieces are placed in a barrel filled with combustibles.

Colorful elephants, giraffes, cats, pigs, lizards, birds and a host of other whimsical creatures are sighted frequently at Galeria el Jardin, Cuyamungue, New Mexico. This mischievous menagerie of raku-fired figures is the creation of artist CJ Buckner of Los Alamos. "Each creature is my interpretation of an animal with a unique personality," she explains.

Although she has made many cats, for example, each has a distinct appearance and is given an individual name, often using a play on words that reflects her acknowledged "off-center sense of humor." One feline is the fiendish-looking "Lucy Fur."

"Basically, my aim is to make each piece fun," says Buckner. "I want the viewer to smile and have an immediate friendly relationship with the critter. I think most people feel that they are adopting a friend, not just buying a hunk of clay. It always amazes me when I talk with people who have one (or more) of my pieces and they refer to it by the name I have given it, rather than as 'one of your cats.'"

Buckner spent her early childhood on a farm, where animals caught her fancy at a young age. That interest in animals and life sciences led to a bachelor's degree in biological sciences from the University of

"Red Neck Chick" (left), and "Red Neck Chicken," handbuilt.

Iowa, and a master's degree in zoology from Arizona State University. Then she decided to follow an urge to try her hand at watercolor painting. She took a few painting classes initially, but was largely self-taught. Since she was living in the high mesa and canyon country of the Pajarito Plateau adjacent to the Jemez Mountains, it was natural for Buckner to paint outdoor scenes of northern New Mexico's grand landscapes and skies.

She worked exclusively in that medium for the next 20 years, selling her landscapes and still-life paintings through regional galleries and art shows. Then came the day when a close friend invited Buckner to join her in Roger Sweet's sculpture class at the University of New Mexico in Los Alamos. Buckner recalls, "I was soon having a good time building animals from clay."

While developing her clay sculpture skills, she continued painting at a reduced level. But after taking a class in raku firing, Buckner realized she needed to choose between painting or clay sculpture. "I decided to pick the medium that was the most fun for me," Buckner says.

Process

She begins each animal's torso by molding a commercial raku clay over a smooth river rock that has been covered with a cloth to prevent the clay from sticking. She covers the cloth with an approximately $^3/_8$-inch-thick slab, which she pats until it conforms to the rock's contours. Trimming excess clay from the "equator," she lets this hemisphere stand until the clay is self-supporting, then repeats the process for the other half. Joining the two halves together results in a form closely resembling the rock's size and shape.

At this stage, the clay will remain workable for several days if stored in a heavy plastic bag. This allows time for her to shape the head, arms, legs and any other appendages by hand and attach them to the torso.

When the clay has thoroughly air dried, Buckner bisque fires the animal to cone 06 in an electric kiln. She then patterns the surfaces of the bisqueware by penciling in lines to define the regions that will receive the various glaze colors. Next, she brushes a wax over these pencil lines. The waxed lines will be black

Left to right: "Purrina Cat," "Old Biddy Hen," "High Minded Giraffe," "Son of a Bun Rabbit," "Mouse Trap Cat," to approximately 8 inches in height, handbuilt, brushed with glazes, by CJ Buckner.

at the completion of the postfiring smoking, serving as both design elements and separators between the different glaze colors.

Once the animals are glazed, Buckner carefully inspects and organizes them for the raku firing. She usually fires several pieces with similar glazes together in a batch, since these glazes will mature at approximately the same time.

Buckner built her raku kiln from a heavy-duty, 30-gallon, galvanized trash can. The can's inside vertical surfaces are lined with 1 to 2 inches of ceramic fiber having a 2700°F rating. Three hardbrick halves on the bottom support a kiln shelf. A single propane burner at the bottom of the can provides adequate heat.

When the glazes are mature, Buckner (wearing heat-protective clothing) turns off the gas and removes the pieces from the kiln. She then places them in a nest of straw that lines the bottom of a retired Civil Defense water storage can. The hot pieces immediately ignite the straw. Placing a lid on the can creates a reducing atmosphere that consumes oxygen from both the air and the metallic oxide glaze.

After the hot pieces cool for about 20 minutes, Buckner quenches them with a water spray. When they are completely cooled, she cleans each with a soft-scrub pad and toothbrush, removing the characteristic residue.

Floral Imagery on Raku Fans

by Marcia Jestaedt

"Bird of Paradise Fan," 26 inches wide, slab-formed tiles, by Marcia Jestaedt.

Producing raku fans with floral motifs is not something I deliberately set out to do, but rather something that has evolved over the past few years. I began with tile fans that were somewhat abstract in design, then moved on to a traditional fan shape decorated with impressed leaves and flowers. I became quite intrigued with the thought of using floral motifs, but knew that pressing actual flowers was not the answer. Aside from the fact that the flowers lost a great deal of their original appearance and definition when pressed with a rolling pin, there is no way to show different views and angles.

To continue in this direction, I had to work with drawings. I have been drawing since I was a very young child; however, I had never studied flowers and am not a gardener. So I began looking at and collecting books, photographs, seed catalogs, greeting cards—anything with flowers on it. I also sketched or photographed flowers at botanical gardens and at a friend's garden.

Process

Each fan image now begins as a full-scale, full-color drawing. It is necessary for it to be complete so that I can refer to it for proper shadings when glazing. The drawing is then placed upside down on a light table and the reverse image, excluding the shading, traced on the back. A sheet of newsprint is then placed on top of this reverse image and the outline is once again traced. This separate reverse image is used to transfer the drawing to the clay.

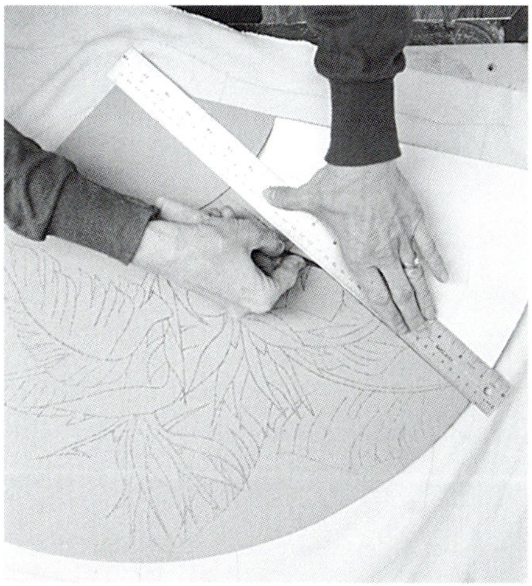

A pencil image on newsprint is inked by going over it with a soluble-ink pen, then transferred to the clay slab by rubbing gently (left). Templates are used to cut the slab into five tiles (right).

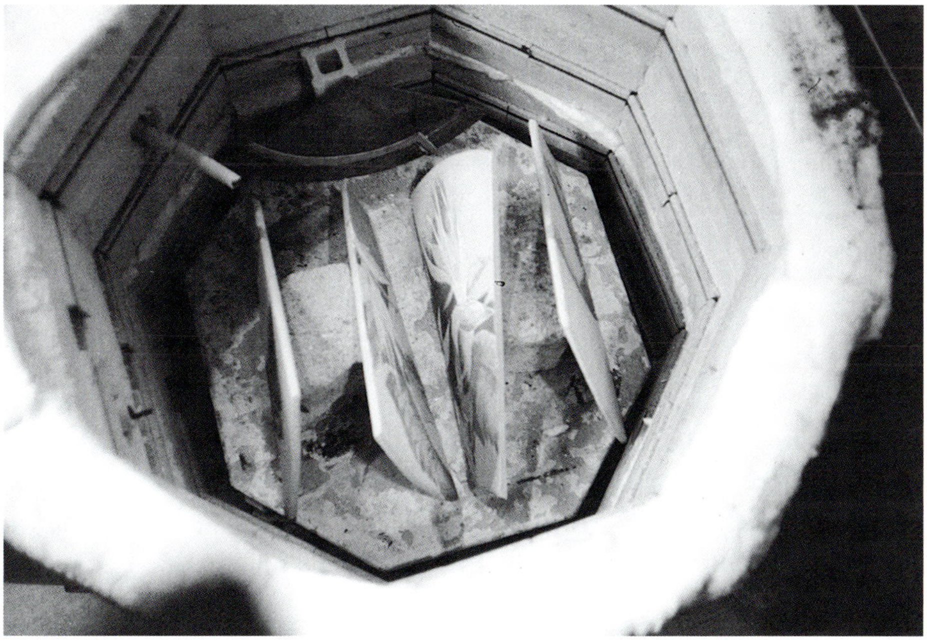

Commercial cone 06–04 glazes are applied in a painterly manner, and the tiles are fired on edge in an electric kiln.

I then roll out the clay and, using a template, cut out the fan shape. Next, the newsprint image is inked by going over it with a soluble ink pen. The inked drawing is placed upside down on the clay and gently rubbed. I work across the drawing, checking as I go, until the ink has been completely transferred.

Templates are then used to cut the fan into five tiles. The tiles are set aside until almost leather hard.

At this point the edges are smoothed and the drawing is incised with a sgraffito tool; any additional carving, surface texturing and sprigging are also done.

Since most of my fans have sprigging on them, they cannot be dried between plasterboard. Instead, I place the five pieces on plasterboard, weight down all the edges with commercial tile shards to prevent warping, and cover with plastic sheeting. After about three days, the pieces are transferred to a slatted wooden shelf, and again weighted down and covered with plastic.

After a bisque firing, the image is redrawn with pencil in the sgraffito lines in order to facilitate glazing. Commercial cone 06–04 glazes are used because of the wide choice available and consistency in colors. They are applied in a painterly manner, combining several colors to achieve the depth and shading needed to give the flowers and leaves a three-dimensional quality.

After glazing, the tiles are fired on edge in an electric kiln to 1850°F, then removed for postfire reduction in birch leaves. Finally, they are attached to ¼-inch birch plywood with a commercial floor- and wall-tile adhesive.

I have found flowers to be an endless source of inspiration and am challenged by the re-creation of their images with clay, glazes and fire.

New Beginnings

by Beth Cavener Stichter

"Ananse's Jar," 27 inches in height, raku-fired stoneware with stains, inks, and acrylic, by Beth Cavener Stichter.

Often, one of the first questions people ask me is "Where did all this come from?" I have a reservoir of clever answers, but the truth is the two people who raised me have an incredible appreciation for the natural world, on the most minute and grandiose scales. My mother, Nancy Jacobsohn, is a sculptor; she taught me the language of clay (which I sometimes feel I speak more fluently than English), and fueled my desire to communicate my ideas with my hands. My father is a molecular biologist; he and I spent endless hours staring at the night sky while he stretched the seams of my imagination with tales of recombinant DNA and evolutionary battles on the microscopic scale.

I spent the first 21 years of my life studying science, working summers in my father's lab and paving the way to Ph.D.-dom. Toward the end of college, my astronomy and physics studies disintegrated into numbers and pages of equations. I was a long way from the ghostly touch of awe that had led me down this path. It was then that my first language resurfaced, and I found my way back to insect exoskeletons and neutron stars through my artwork. Then I received the staggering news of my parents' impending divorce. Little did I know the ending of their relationship would act as a catalyst

"Horse God," 16 inches in height, handbuilt, raku-fired stoneware, by Nancy Jacobsohn.

for my mother's and my own artistic careers, as well as signal the beginning of our unique friendship.

I spent the following year studying in Florence, Italy. Upon my return, I apprenticed with sculptor Alan LeQuire in Nashville, where I learned about mold making and the bronze casting process. After graduating with a B.A. from Haverford College in Pennsylvania, I spent an intense semester building my portfolio at the Appalachian Center for Crafts in Tennessee, trying to discover who I was as an artist.

Prior to my time at the craft center, my education had involved a very classical approach to the fine arts. I had concentrated on honing my skills on the human figure, while ignoring the creative side of my work. When I faced the challenge of building a portfolio to approach galleries, I realized my work lacked coherence and didn't truly reflect any part of myself. I had rediscovered the language of clay, but needed to figure out what I wanted to say.

At that point, my ideas were scattered in so many directions, I spent more time staring at blank pages in my sketchbook than working in clay. I was fighting the age-old dilemma of craft versus fine art. Throughout my four years at Haverford, I had been taught that clay was simply an intermediary material, to be used only for practice or as a prelude to bronze casting. But I had grown up with a love for the material itself. Stepping across that fine line, I feared the craft world would not embrace my ideas as a ceramics sculptor. I chose to walk along the line itself, as did my mother, each of us leaning toward a different side.

Throughout my teenage years, my mother and I lived miles apart in the same house. The only gap in the stone wall we had built between us was a shared love for horses. By

"Gone," 15 inches in height, stoneware with white crackle glaze, raku fired, by Nancy Jacobsohn.

"Giraffe," 19 inches in height, glazed stoneware with stained beads, raku fired, by Nancy Jacobsohn.

the time I was 13 and going through a severe emotional depression, my only solace was the horse I had saved for since I was 6 years old. Here is where my mother and I found common ground. We made an uneasy alliance and somehow managed to buy two horses, supporting them with second jobs and pinched pennies. In the difficult years before I left for college, the horses somehow kept our tenuous relationship together.

During my junior year, I learned not only that my parents were about to be divorced, but my mother also required a potentially dangerous back operation as a result of childhood polio. The doctors advised her to sell the horses, claiming she would no longer be capable of riding or taking care of them. It seemed like a bitter end to the dream that both she and I had worked to achieve. To the doctor's amazement, she fought her way

Beth Cavener Stichter and her mother, Nancy Jacobsohn, preparing to remove a piece from the kiln for postfiring reduction.

to a full recovery from the operation, and was eventually able to bring the horses back into her life.

As time passed and we put our lives back together, the two of us forged a stronger relationship. In the last three years, we have each taken the giant leap into the unknown. My mother, starting a new business of her own, and myself, fresh from college, have set out on similar paths of discovery of ourselves and our work. It has been an exhilarating and unnerving experience that has brought the two of us together, and has provided us both with the necessary tools to make our careers a reality.

About the Work

My own investigation has centered on three different means of expression. I chose to work with insects because of the fascinating mixture of fear, revulsion and awe that they inspire in most everyone. I created my own mythology in which strange

The piece is nestled in newspaper and other combustibles, then a horse watering trough positioned over all.

alien creatures disguise themselves with our own human nature and culture. There is a long tradition of insect representation in mythology, superstition and folklore, which has provided me with an endless source of dark and whimsical images.

It seems only natural that another frequently occurring subject in my work is the hand. Aside from our faces, our hands are one of the most expressive features of the human body. Lastly, I have found simple grace in the visual rhythms of muscle, bone and sinew, and these elements have woven themselves into many of my pieces. Through this formalistic abstraction of organic, mechanical and subconscious forms, I can reexamine the everyday world, suggesting a different way of looking that will encourage viewers to look closely to see what is there.

My mother's work has evolved from her passion for horses. During the recovery period after her back surgery, she was in a great deal of pain and very discouraged, so the doctor urged her to spend as much time in the studio as she could. The limit then was about 15 minutes, and all she could make were beads. So she made piles and piles of them. When my mother was able to sculpt again, she made a horse with a beaded mane and tail that she called "The Horse God." It wasn't that she thought it had any supernatural power, but it was invested with all her pain, regrets and unrealized dreams. She has made many horse gods since then, and tells customers that each comes with three wishes. While it isn't guaranteed that he will grant them, she does guarantee he will listen and consider them.

"Vessel #2," 20 inches in height, stoneware, with brown glaze and black stain streaks, by Beth Cavener Stichter.

The bodies of her horses and other quadrupeds are quite massive in relation to their long thin legs, so they are coil built from the legs up over metal-rod armatures. The rods are held upright by bats with holes drilled through; once the piece is leather hard, the rods are pulled out through the bat.

Recipes

Dry Alligator Rust Raku Glaze

Bone Ash	2 parts
Gerstley Borate	4
Nepheline Syenite	1
Copper Carbonate	1
	8 parts

Jacobsohn "inherited" this recipe from Nancy Humeniuk. She uses a mouth sprayer for a thin application. When applied thickly, it produces an interesting cratered surface.

Vince Pitelka's Smooth Matt Glaze

Cone 5–8

Dolomite	18 %
Whiting	4
Nepheline Syenite	17
Potash Feldspar	40
Ball Clay	11
Kaolin	10
	100 %

This clear glaze is applied over the surface of slips and stains to give depth of color. It is usually sprayed on lightly with a mouth sprayer.

Clear Soldner Raku Glaze

Gerstley Borate	80 %
Nepheline Syenite	20
	100 %

This is a tried-and-true clear crackle glaze with a horsehair brush.

Vince Pitelka's Raku Body

Cone 8–10

Ball Clay	25 %
Cedar Heights Goldart	25
Fireclay	25
Fine Grog	25
	100 %

Beth Cavener Stichter uses this clay body for most of her large-scale pieces. It builds very quickly, and holds up well against cracking or warping.

Production Raku Kiln

by Ronda M. Liskey

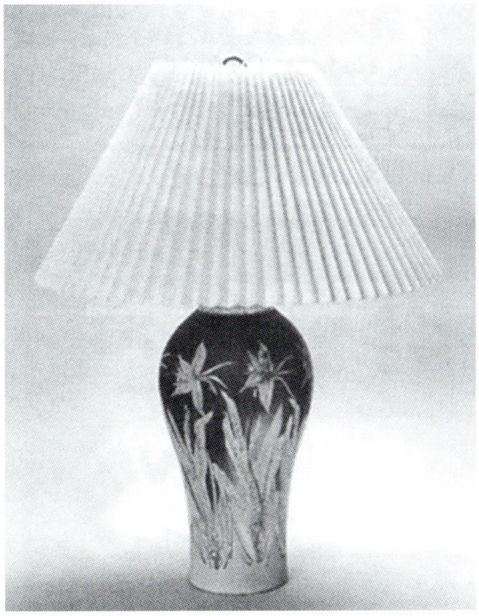

Lamp with raku-fired base, 26 inches in height (including shade), wheel-thrown stoneware, with lusters, reduced in pine needles, black walnut leaves and straw, by Ronda M. Liskey

I enjoy most everything about raku firing. The suspense of opening a kiln at 2000°F, and seeing the pots almost breathe with a searing translucence, is a remarkable experience. Then the added straw and pine needles create a fire that dances around the pots, painting a unique pattern on each one. Some call it instant gratification or cheap thrills; I say it's just delightful.

Safety, quantity production, and ease of firing guided the design of this two-chamber raku firing system.

Firing and reduction take place on 3½-foot-square pads of firebrick, laid (level) one brick deep on the ground.

Fulcrum posts set between the firing pads allow the chambers to be levered from one pad to the other.

At 2000°F, the kiln chamber is lifted, and the ware covered with combustibles.

The drumlike raku chamber can be easily raised and lowered to check postfiring reduction.

But, while offering many otherwise unobtainable results, rakuing can involve hazards, inconveniences, and even injuries. My kiln system was designed to address both production needs and safety concerns.

Construction

Built for about $2000 from new materials, this system features two firing pads where pots are both fired and reduced. The kiln and reduction chambers are each suspended from a steel pole that is anchored on a swivel base atop a 4x4 post. The pole acts as a lever (counterbalanced by a 50-pound weight), easily moving the drum-shaped chamber up and down and from side to side.

Firing

After one of the pads is loaded with glazed ware, the kiln chamber is lowered over the pots. Guide poles

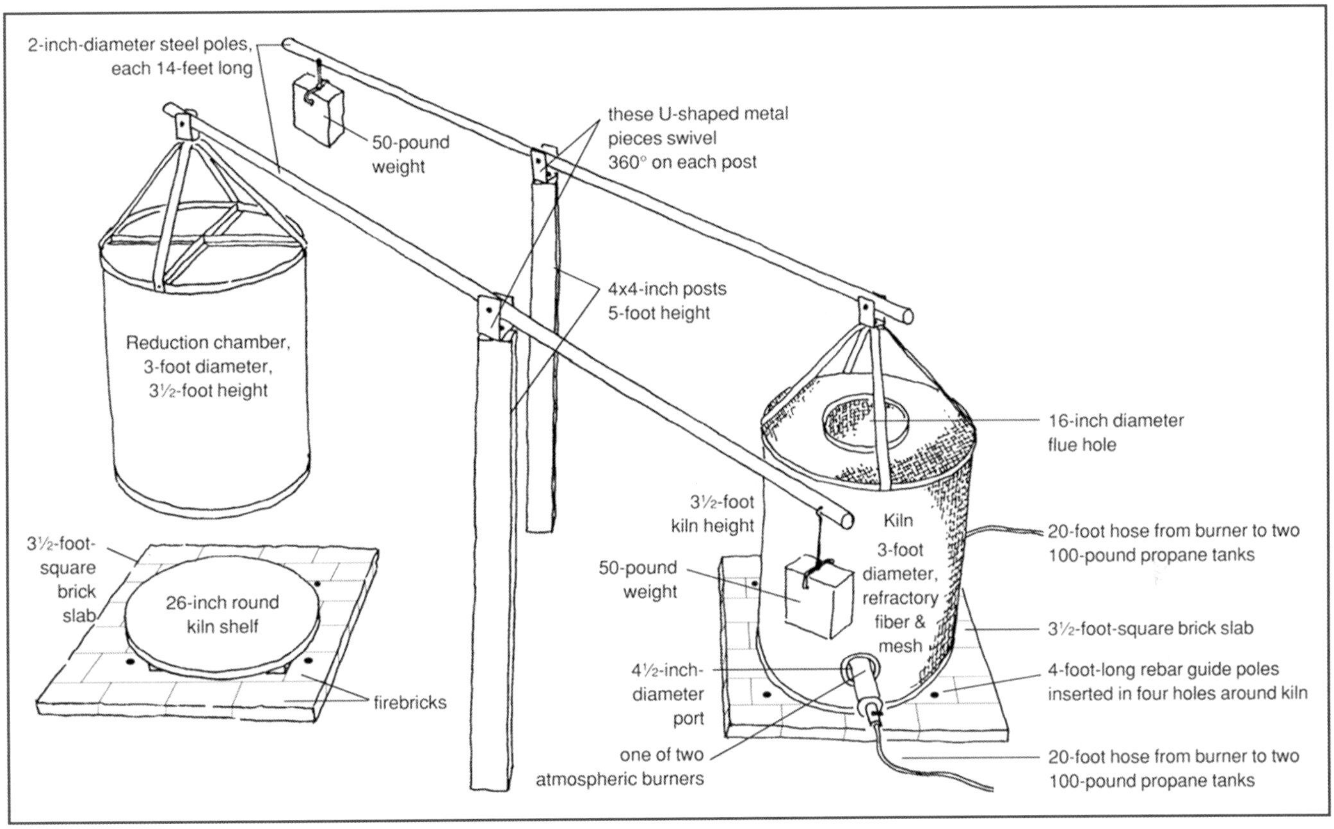

help to center the kiln. It is heated by two venturi burners, each hooked up to two 100-pound propane cylinders linked together with a 35-pound regulator. The temperature smoothly and efficiently climbs to 1950°-2000°F in 30 or 40 minutes. Then the burners are removed, and the kiln chamber lifted and swung aside to cool. Immediately, bundles of pine needles and straw or leaves are placed on and between the hot pots, and the reduction chamber is quickly lowered over all.

While the first load smolders, the other kiln pad is loaded with pots and topped with the kiln chamber so that the process can be repeated. Thirty minutes later the reduction chamber is emptied of the first firing's bounty so that it is ready to cap off the second load when pots reach temperature. Then the first kiln pad is loaded for a new (third) firing, and so on. When all the firings are completed, the drums are easily detached and stored with the other kiln accessories.

Safety

Five safety precautions are always followed when using this system:

1. Before beginning firing, the area around the kiln site is watered to prevent a grass fire; and there is always a water hose ready for action.
2. The propane tanks are located at least 15 feet away from the kiln and the burners are taken well away from the kiln before the hot chamber is lifted.
3. Instead of peering into the kiln to check the glazes, and thereby breathing potentially toxic fumes and exposing my retinas to harmful radiant heat and light, a pyrometer is used to determine readiness.
4. During firing, protective head covering, boots, heavy clothing, insulated elbow-length gloves, goggles and a face mask are worn to avoid burns and excessive smoke inhalation.
5. Finally, moving the reduction chamber to the pots, instead of using tongs to move the heavy pots to reduction cans, prevents back strain.

Besides eliminating many risks, this removable-chamber system allows me to raku fire in two hours what once took an entire day with one small kiln. Although some reduction control is sacrificed, there are new and exciting results, which I am learning to master for my own purposes.

Alternative Raku Kilns

by Andreas Salzman with Jennifer Salzman

"Bridge," 100 inches in length, intaglio on clay, wood and leather.

My first kiln was nothing more than a 55-gallon drum barrel. Inside the barrel, I balanced a pot on top of a few bricks, and filled the remaining space with wood, paper and anything else I could find that would burn. The pot came out amazing—cracked and ugly, but amazing. Shortly after this experience, I graduated to a pile of broken hardbrick and a natural gas burner. I still made a lot of ugly broken pots, but continued to gain more and more enthusiasm for unpredictable results.

I then spent a few years raku firing with a propane-fueled softbrick kiln. This allowed me to develop a little more consistency through a better understanding of glazes and firing principles. As I developed more proficiency in the raku process, my results became more predictable. While consistency is an important tool, it can turn a once-exciting process into a chore.

In graduate school, my work took a dramatic turn away from ceramics to mixed-media sculpture. I was excited about different materials and how I could manipulate them. Everything became an experiment. Then I decided I needed to develop a broader palette for the raku elements that made up a small but important aspect of this work.

To explore possibilities in raku techniques, I focused on the kiln design because I felt I had done enough work with glazes and postfire reduction. I decided I would try to manipulate the atmosphere of the kiln by firing with solid fuels and vapor glazing. The very first steps of what I am sure will be a long journey follow.

Building a Saggar Kiln

The first design was a saggar kiln. Based on a more traditional raku kiln design, it was simply a round barrel, about 16 inches in diameter, made of castable refractory and Fiberfrax. I began by building a form from cardboard, duct tape, polystyrene insulation and plywood bats—all materials I had on hand. Two 16-inch bats served as the top and bottom of the cardboard form. A 30-inch sheet of cardboard was stapled to the bats and the whole thing covered with duct tape (to waterproof it). The bottom bat was then attached with wood screws to a 24×24-inch plywood base.

Next, I made a second cardboard tube slightly larger than the first; 1-inch cuts every 2 inches around the bottom formed tabs. After covering the second tube with duct tape, I placed it over the first tube, then stapled the tabs to the plywood base. The space between the two tubes, when filled with castable, would become the 2-inch-thick wall of the kiln.

The last step in building the form was to cut three cardboard strips to act as dividers during the casting, separating the wall into three vertical sections. This would allow me to dismantle the kiln, making it relatively portable.

The castable refractory was mixed with slightly more water (with no apparent adverse effects) than according to the directions provided by the manufacturer. I knew more water could weaken it, but the alternative was castable that would be too dry for me to work with. I have fired it several times and it seems fine.

I poured the castable into the form a quart at a time, tamping (ramming) it with a piece of scrap wood. When the form was filled to about 10 inches from the top, I slid in three 3×5-inch polystyrene spacers to leave three ports along the bottom of the kiln (being cast upside down), then finished pouring and ramming the rest of the castable.

For the floor of the kiln, a castable slab was covered with a layer of Fiberfrax (more for a tight seal than insulation). The lid was made from fiber with a flue opening cut in the center.

The ware itself was fired inside a saggar thrown from a heavily grogged raku clay, with a cover that had a 2½-inch peephole in the center. The size was determined by the need to allow at least 2 inches of space between the saggar and the wall of the kiln. For reinforcement, I wrapped the saggar with wire, expecting it would crack during the firing (which it did).

None of the saggars I made lasted beyond one night of firing. Even though wired with Nichrome, they fell apart when I removed them from the kiln. I tried many variations in design, but they all cracked badly.

Wood-fired raku kiln with a castable ware chamber, and a fiber door and lid.

What I did discover was that some changes in the saggar design seemed to have a great effect on the ware. If I perforated the saggar, greater amounts of "crud" would get into the setting, affecting the glazes. Tighter saggars heated slower and tended to promote stronger reduction.

A portable saggar kiln, made from castable refractory sections and fiber insulation, was fired with a mixture of wood and coke.

I placed the saggar inside the kiln on a stack of bricks about 10 inches high. The space below was packed with a mixture of wood and coke. The 2-inch space between the wall of the kiln and the saggar was also packed with fuel. Finally, two of the three ports were sealed off with fiber before the fire was lit.

Firing Cycle

The coke did not burn easily. I had to have a pretty good fire going with the wood before the coke started to burn. After a while, I could use a blower to really get things going. In later firings, I became impatient and ended up using the propane burner from my gas kiln to light the coke. That seemed to be the best solution. After ten minutes of gas, I fired up the blower to quickly increase the temperature. I was amazed at how hot the coke burned.

When the saggar started to show color, I used tongs to position the first piece (a very thin layer of grog or sand in the bottom of the saggar helps to reduce the chance of breaking the ware), then placed the lid on the saggar and covered it with more fuel (making sure that the ware could still be seen through the peephole).

When the glaze matured, the lid of the saggar was removed, and the ware was pulled for smoking in the standard American raku tradition, or allowed to cool in the Japanese tradition. The next piece was then placed in the hot saggar and the cycle repeated.

After seeing how well the coke burned, I tried to eliminate the use of wood altogether; however, with coke alone, the glazes tended to blister and scum (I think this could have been from the sulfur in the coke). Using wood kept the flames longer and seemed to eliminate most of the glaze problems.

I had some trouble getting a good mix of wood and coke until I discovered the wood pellets used in pellet-fueled wood stoves (available at home centers and hardware stores). They come in a variety of wood types, and cost a few dollars for a 40 lb bag.

Fuming

I like to take advantage of the confined atmosphere and retained heat of the saggar to "fume" the work before it is pulled. Many chemicals and combinations work, but my favorite is 10 to 50 grams of a 50% sodium bicarbonate/50% copper carbonate mixture sprinkled into the saggar. I replace the lid of the saggar and give it a few minutes to volatilize. Of course, this usually makes a mess of the saggar.

Obviously, there are some serious safety concerns regarding fuming. The most difficult part is putting the materials into the saggar without poisoning or burning oneself. I use a respirator fitted with a vapor filter (and still hold my breath), as well as a face shield and gloves, and I always stand upwind from the kiln as the fumes clear. My biggest concern is the safety of spectators. I make sure everyone present takes appropriate precautions.

A somewhat safer method is to place a mixture of salt and metallic oxides in the bottom of the saggar before the fire is lit. The bottom of the ware can be protected by placing it on a small piece of softbrick. The disadvantage is that the effects become less intense in successive cycles.

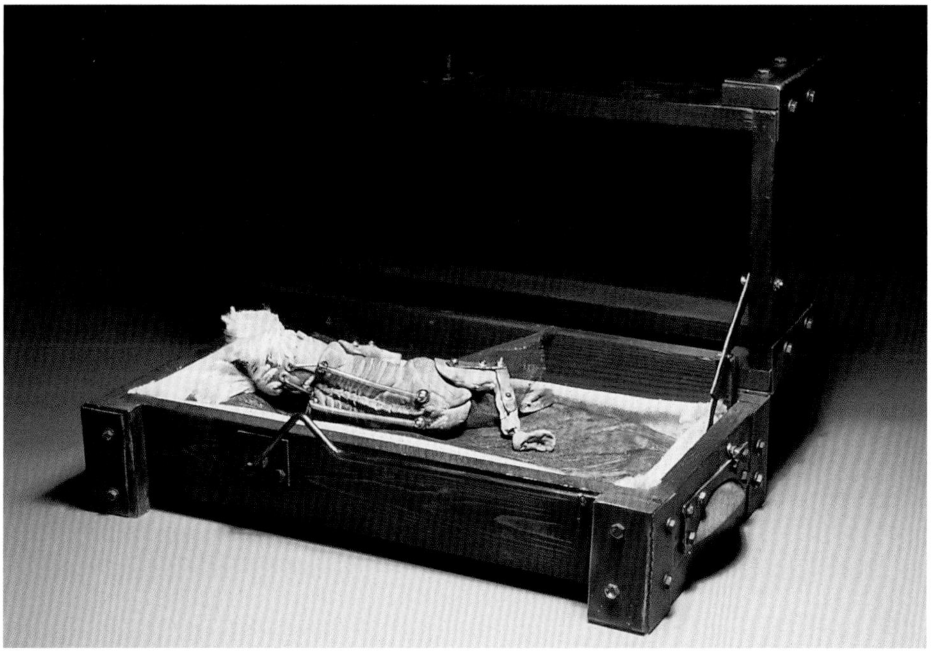

"Father's Hopechest #2," 21 inches in length, wood, zinc and raku-fired clay, by Andreas Salzman.

Building a Wood Kiln

While the saggar kiln worked pretty well, it did have some drawbacks: the heat transfer through the saggar to the ware could take a long time; the saggar itself limited the size of the ware; and stoking could be difficult. With this in mind, I developed a larger and slightly more involved design that would eliminate the use of coke, and fire completely with wood. The coke had worked well in the small confined space of the saggar kiln, but I suspected the longer flames and wood ash might have a positive effect on the glazes. I also hoped I could experiment with some salt glazing, and thought that an open firebox and ware chamber would work well.

The new design had a much larger firebox and a combustion chamber, which would allow firing to a much higher temperature. I didn't intend to change the maturing temperature of the raku glazes, but rather shorten the firing cycle, and thus increase the practicality of a small kiln. The higher temperature range would also give me more flexibility in regard to experimental firings, maybe even reaching stoneware temperatures.

There was no saggar in the new kiln design. Instead, the combustion chamber served as a foundation that could be fitted with a variety of ware chambers. This allowed me to continue to experiment with the designs to fit other firing temperatures, atmospheres, ware sizes or whatever else I came up with.

Again, I decided to use castable refractory. I liked the idea of making the foundation chamber out of as few parts as possible, and it would be less likely to shift or move around when it was being stoked. It would also be lighter and less difficult to transport. Most importantly, I had a greater supply of castable than I had of hardbrick. In a perfect world, I think that hardbrick would be the better choice because of versatility and ease of construction.

I basically followed the same steps to cast and ram the new firebox/combustion chamber as I had on the saggar kiln. I started with a sheet of plywood and worked as if I were building the kiln upside down. A polystyrene form was attached to the plywood. I then used more plywood to form the outer walls. To ensure that the castable would set properly, all the plywood was sealed.

The plan was to cast a firebox that would measure 11 inches wide, 11 inches deep and 22 inches long. This would lead into a 20-inch-square, 11-inch-deep combustion chamber.

The firebox/combustion chamber was divided into four pieces, each weighing about 50 or 60 pounds. This would be heavy enough to remain stationary during the firing, and yet light enough to be lifted.

The final element was a collar of castable that would serve as the transition point from the square combustion chamber to the round ware chamber. All five elements were then assembled with fiber gaskets at all of the joints.

The first ware chamber was made from fiber and hardware cloth. This was later replaced with a castable chamber with a side door made of fiber. By that time, I had found that using cement deck footing forms worked much better than the duct tape and cardboard I had used for the saggar kiln. Available at building-supply stores in various diameters up to 6 feet in length, these forms are made of heavy cardboard covered with wax and can be cut with a hand saw. To decrease the diameter, I cut the interior section vertically and overlapped the seam. Polystyrene spacers were again used to create the ports.

Once everything was cast, I assembled the ware chamber with metal clamps made from a 1¼-inch by ⅛-inch steel bar. Castable has a habit of cracking at transition points, and the use of metal clamps helps to lessen this.

Firing Cycle

To fire, I placed a metal grate in the bottom and a kiln shelf along the top of the firebox extending over the top of the combustion chamber (the kiln shelf was narrow enough to allow flames to pass into the ware chamber), and used wood that was as dry as possible, split into small kindling-size pieces. I judged the stoking rate by keeping a reasonable-sized flame in the flue opening. One of my early concerns was that the flame length might be difficult to control, making the temperature difficult to regulate. Through experimentation, I found that removing part of the kiln shelf cover from the firebox allowed me more control over the flame length. I also found that I could take advantage of any wind by angling bricks into a funnel directed toward the firebox.

Once I had a good bed of coals, I started the blower. By using a blower, I could control the flame length and greatly increase the temperature (it was possible to go from 1500°F to 2200°F in less than 20 minutes). To reduce the mature glazes, I turned the air down, stuffed the firebox and closed the damper. I had to watch the pyrometer closely to make sure the temperature stayed above 2000°F, which allowed the glazes to remain fluid. After 10 minutes or so, the ware was ready to pull.

Vapor Glazing

When salting, I leave my work unglazed, covered only with a thin stain of ferrous sulfate or copper carbonate mixed with 10% borax. The reduction cycle was the best time to introduce salt for vapor glazing. I threw about 3 or 4 ounces of salt into the firebox every 3 minutes. Two or three saltings were usually enough for a thin glaze. After the last salting, I turned down the blower and started the reduction cycle.

Fuming was done at this time as well. The best results were achieved by fuming after the reduction cycle. The door of the chamber was removed, and fuming agents (such as silver nitrate and stannous chloride) sprayed directly on the ware or shelf. The door was then replaced and the fire stoked another 4 or 5 minutes.

For me, both kiln designs hold many possibilities worthy of further experimentation. The combination of glaze chemistry, fuming, fuel residue and postfire reduction seems to offer an inexhaustible number of variables to explore. After all, the vast possibilities are what first drew me to raku.

Wood-fired Raku Kiln

by Nesrin During

Handbuilt vessel, approximately 6 inches in height, with iron-spotted raku glaze, reduced in sawdust, by Nesrin During.

Texel is a small Dutch island on the North Sea. It is an island of sand dunes, natural lakes, sheep, tourists, wind and rain. In these quiet surroundings, I have been making ceramics for over 20 years. Some of my work is wood fired in a simple, self-built raku kiln. It is simple because it is actually made of stacked bricks and sized to fit what I'm planning to fire (sometimes I start firing small objects, then aggrandize the kiln by adding another one or two rows of bricks for the bigger objects).

These kilns are built wherever I want them to stand, taking into consideration the direction of the wind. The day after I've finished firing, I put away the bricks, out of sight under a roof or in a shed to keep them dry until the next firing.

Kiln Construction

A typical kiln is built from about 50 insulating bricks (IFBs), a square kiln shelf (size depending on what I want to fire), a metal grill for a grate, a piece of sheet metal, and some broken shelf pieces. It takes about 45 minutes to build, and fires to about 1650°F in about 45 to 60 minutes for the first firing; thereafter, every load takes 15 to 20 minutes (one can see

Wood-fired raku vessel, approximately 6 inches in height, handbuilt, with poured glazes.

the glaze melting, the pots shining in the flames, from the top of the kiln).

To begin, I level the surface with a layer of dirt; it will also protect what is beneath (concrete, for example, can crack with the heat). I look at the direction of the wind and accordingly build the firemouth to receive the wind.

Depending on the size of the kiln shelf (it's going to diagonally span the walls), I lay two courses of bricks to establish the back and side walls. Upon these the grate is placed so that the ashes can fall through (figure 1). (My iron grate was salvaged from a dump; one could also use a kitchen stove grate, which would last a few firings). Then another two rows of bricks are laid, and the kiln shelf placed diagonally so that the walls support three corners. A piece of sheet iron (which was also salvaged from the dump; one can also use a piece of angle-iron) across the front of the kiln supports the fourth corner (figure 2).

Because some of the bricks in the following course will stand on the kiln shelf, I have to raise the others to the same height with pieces of broken shelves. After this leveling course, a couple of rows are laid in a diminishing circular pattern

Two courses of bricks are laid to establish the back and side walls, then an iron grate positioned on top.

After another course of bricks, a sheet of iron across the firemouth supports the corner of a kiln shelf.

Some of the bricks in the following course stand on the kiln shelf, so level with pieces of broken kiln shelf.

The kiln fires to about 1650°F in 45 to 60 minutes, while pieces for the next firing preheat on top.

(figure 3). I also partially close the top of the kiln with broken shelves, leaving a central hole to function as a chimney.

Firing

I fire this kiln (alone or with the help of students) with scrap wood. The wood should be dry and thinly split. I start with some newspaper and thinly split wood (figure 4).

The fire is well fed in the beginning in order to get the bricks to temperature, but after the first load is done, I fire with less wood and rake the fallen ash from underneath the grate to ensure the fire has enough oxygen to burn well.

Because the kiln is so loosely stacked, I can easily create new openings in between bricks to manipulate the flames toward a certain direction, causing interesting effects on pots.

The pots intended to go into the next firing are placed on top of the kiln to dry. When the glaze on

When the glaze on the pots inside is molten, remove a few bricks from one side of the chamber to facilitate removal of the pots for postfire reduction.

Nesrin During removing a pot from a wood-fired raku kiln loosely constructed from brick.

the pots inside the kiln is shiny and molten, I remove a few bricks from one side of the top to facilitate taking the pots out (figure 5); the pots are then placed in a metal bin containing sawdust (figure 6).

The glaze is a simple: alkali-borate frit (70%) with kaolin (30%). With additions of 0.5% to 3% iron oxide, this gives very beautiful pinks and grays (pink to pomegranate red in oxidation, and gray to black in reduction). Additions of 2% to 3% copper carbonate yield apple green in oxidation and bordeaux red in reduction. Other oxide combinations, such as iron and copper, or iron and manganese, are also good.

The amount of crackle depends on the type of clay used and the handling. After taking the piece out of the kiln, if you keep it in the air for a minute or so before putting it in the reduction container, you're bound to get more crazing, especially on a smooth surface. If the surface is rough, you're more likely to get iron oxide "bleeding," producing thousands of little dots and giving the piece a rocklike appearance.

Building such a kiln requires few materials, and fuel for firing is equally economical. Because it is a wood firing, there is bound to be some oxidation and some reduction at the same time. The resulting variation in surfaces gives beauty and individuality to the work. Not every piece comes out well, but some are really wonderful.

Wood Firing Basics

by W. Lowell Baker

Wood burns in two distinctly different stages. The first, and most obvious, is the burning of gasses produced when wood is heated. Wood begins to gasify at about 500°F. The second is the burning of the charcoal. This happens, for the most part, after the materials that form the gasses have been driven out of the wood. The coals in your ash pit serve to provide some heat to the kiln and to gasify the freshly stoked wood, mostly through radiant heat energy. As the gasses burn in a wood kiln, they typically produce very long flames. These flames can be easily over 30 feet long. Charcoal produces very hot, but very short, flames. The flame from charcoal is normally less than a few inches long. All of these issues are relevant to building and firing any wood-burning kiln.

One of the demonstrations I take my students through when we begin talking about kiln design is to bring an oxygen-acetylene torch into the classroom. If the torch is ignited with only acetylene (fuel), it produces a very long, very dirty flame. One can quickly pass his or her hand through this flame without any real danger, but it will be covered with black soot. As oxygen is added, the flame shortens and becomes significantly hotter. As the flame shortens with the changing oxygen-fuel ratio, smaller flame tips appear in the center of the flame. This is the place where the flame is the hottest. The more defined the tips are, the hotter the flame. You want this part of the flame in the firing chamber of a kiln, not in the firebox or the flue.

If you have a small kiln and a fuel that develops a long flame, you need to either redesign your kiln to use the length of the flame, or simply shorten the flame to bring the hottest part of the fire back into the chamber where the pots are stacked. As with the acetylene example, the easiest way to shorten the flame and make it hotter is to add oxygen.

If you have electricity at the kiln site, adding a blower is one of the easiest and most controllable ways of adding oxygen.

A small squirrel-cage fan that will deliver about 100 cubic feet of air per minute will supply all the air you will need to fire a small kiln. You can fabricate a bolt-on connector to attach the pipe to the blower, or duct tape a piece of automotive tail pipe to the blower. You should realize that the end of the metal pipe will be subjected to a great deal of heat and will have to be replaced after a number of firings. Place the pipe in the ash pit of your firebox and adjust the air-input damper to the desired air flow. You will find that the flame around the blow pipe will be very intense. This system will allow you to fire your kiln with a much smaller firebox than would normally be needed in a natural-draft kiln. The smaller firebox will require more frequent stoking, simply because it will not hold as much fuel as a larger box.

Increasing the flue height would be the last choice in a small kiln. If you do this, you must be certain that you have air intake ports and a flue cross section large enough to allow easy circulation of hot gasses. A damper will be essential for control. This will be less responsive than a forced-air system and will vary more due to atmospheric conditions, because it depends on lowered pressure to bring air into the kiln.

So, more air shortens the flame and more air increases the temperature of the coal bed to help gasify your fuel more quickly.

Portable Fiber Raku Kiln

by Daryl Baird

I began raku firing at my local community college using a gutted electric kiln. Later, I put my own kiln together using a metal trash can lined with Kaowool™ Cerablanket®, outfitted with a weed burner and fueled with propane from a barbecue tank. This design was fairly inexpensive, simple to construct, and easy to transport and set up. But, good as it was, this kiln had a several drawbacks.

The time arrived to build a new raku kiln, so my engineer-potter friend, Robin Smith, and I went to work. Our first task was to list what we wanted in the kiln, then we pulled together the best design elements we'd seen in other kilns over the years and made up a list of requirements. The kiln needed to be:

- Low-cost, sturdy, lightweight, efficient and safe
- Made of readily found materials
- Easy to construct using ordinary tools
- Capable of handling a pot 15 inches high
- Easy to access when removing pots
- Safe to look inside during firing
- Portable and easy to set up

Kiln Construction

For a 19-inch diameter kiln, cut 64 inches of fencing material from the roll. This allows for a 4-inch overlap (figure 1). Roll and shape the cylinder, check the diameter, and then use J-clip pliers to install J-clips down the length of the frame (figure 2). Twisted wires also work, but they bend the ends inward.

Carefully flatten the remaining fencing for the top, and trim it to a 20-inch square. Center the piece on top of the kiln, mark lines at a 45° angle just outside the edge of the kiln frame on each corner (figure 3).

Bend the corners to make the lid's "feet." These feet elevate the inside of the lid and protect it from damage when setting the lid on the ground (figure 4).

To provide a nice view of the kiln's interior, install the 6×6 Pyrex window on the overlap six inches down

CAUTION

Wear gloves and dust mask while handling ceramic fiber.

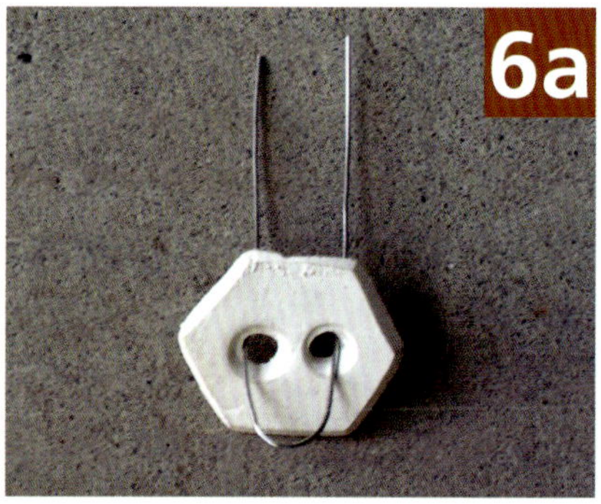

from the top of the frame (figure 5).

Stand the kiln frame on the floor and loosely fit the fiber blanket inside of it. Mark the excess material to be trimmed leaving less than an inch of overlap. Note: The fencing and fiber are both 24 inches wide. If the fiber is a little longer than the fencing, do not trim it. With repeated firings, the fiber will likely settle down for a better fit. Lift the fiber out and lay it on a table, and cut it to length using a sharp knife and the carpenter's square. With the Pyrex window in the 6 o'clock position, reinstall the fiber in the frame with the seam at 3 o'clock. Cut a piece of wire that's 3–4 inches long. Bend it into a U-shape (figure 6a) and push it through the fiber until it's snug against it (figure 6b) and the wire tips protrude through the kiln frame. Hold the button in place with one hand and trim the tips so they protrude only ¼ inch past the frame. Back the wire out and measure its new length, then cut the remaining wires to length.

Put four buttons on one side of the seam, locating the first one an inch from the top and an inch in from the seam. Push the wires through and use the needle nose pliers to bend

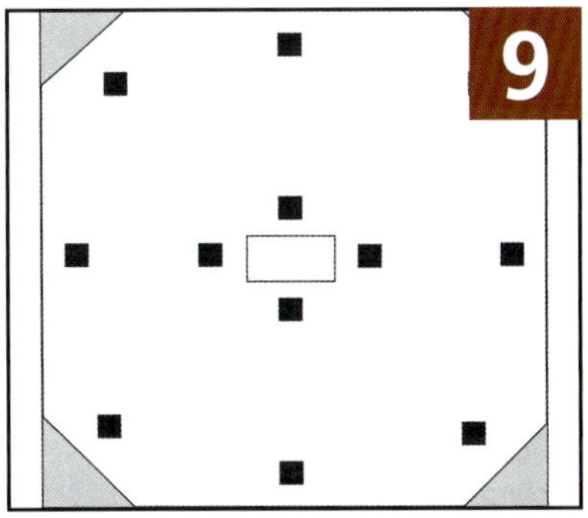

each tip into a tiny hook that attaches to the kiln frame. Install two buttons in the 12 o'clock position—one an inch from the top and the other 8 inches below. At the 9 o'clock position, insert four more buttons. At the 6 o'clock position, place a button an inch from the top, then place one so that its bottom edge is located just above the viewing window (figure 7).

Push a wire into the blanket at each corner of the viewing glass and use the pen to mark where the wire comes through. Draw straight lines to connect the four marks and you will know where to place the button. Later these marks will help you know where to cut out the fiber to reveal the window. The third button in this column goes an inch or two below the window and the last button just about an inch off the bottom. Place a button on either side of the blanket area covering the window. Return to the 3 o'clock position and put in four more buttons down the other side of the seam. For the burner port, cut a 4-inch wide by 5-inch high hole on the side opposite the viewing window. Secure the fiber with a button on each side of the port. Use a half-height button to secure the fiber under the burner port (figure 8).

Before lining the lid, remove the eight center-most rectangles with wire cutters to make a 4×4-inch flue opening. Cut the remaining fiber piece down to 20×20 inches, then press it into the lid frame and trim the excess. Do not cover the upturned corners of the lid frame. Use twelve buttons on the lid, with eight around the perimeter and one on each side of the flue opening (figure 9).

Building a Kiln Cart

This kiln can be placed on a base of insulating fire bricks (ISBs), located on top of a concrete slab or concrete blocks. We wanted portability so Robin designed and constructed a simple cart. This cart makes it easy to move the kiln out of the way, even while it's still very hot. Building the cart requires basic skill with a metal-cutting saw, stick welder, angle grinder and welding clamps.

Construction

Cut four pieces of angle iron (a) to the same length for the ISB frame.

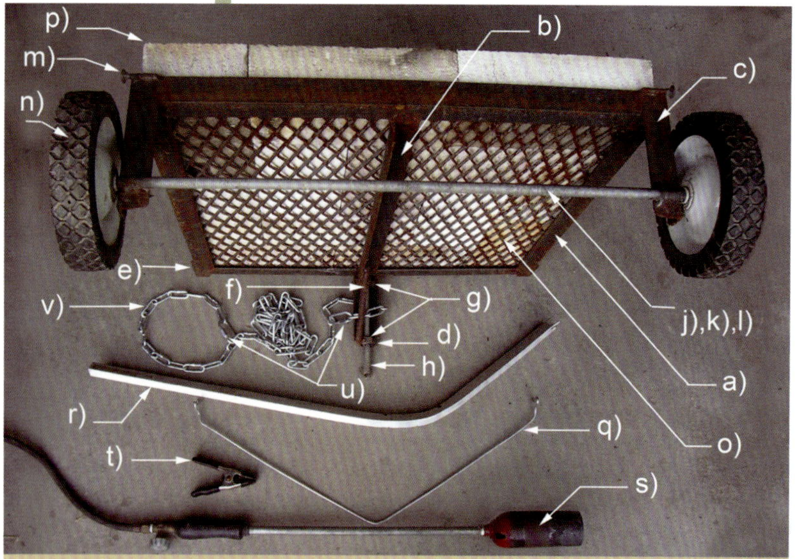

Cart Material List

ID	Description	Material	Qty
a	Frame	⅛×1¼×22¾ inch angle iron	4
b	Centerline Brace	⅛×1¼×20¼ inch angle iron	1
c	Back Leg	⅛×1¼×6 inch angle iron	2
d	Front Leg	⅛×1¼×6 inch angle iron	1
e	Gusset	⅛×1¼×1¼ inch angle iron	2
f	Leg Brace	½×6½-inch-long round stock	1
g	Nut	½ inch diameter	2
h	Leveler	½×9-inch threaded rod	1
j	Axle Rod	½× 26¾-inch threaded rod	1
k	Flat Washer	½ inch flat	2
l	Nut	½ inch	2
m	Carriage Bolt	¼×1¼ inch	2
n	Wheels	7 inch diameter	2
o	Exp. Metal Mesh	22⅜×22⅜ inch	1
p	K23 Softbrick	4½×4½×9 inch	13
q	Mount Wire	9 ga.×31 inch	1
r	Mount Trough	⅛×1×36 inch aluminum angle	1
s	Weed Burner or Raku Burner		1
t	Squeeze Clamp		1
u	Quick Link		1
v	Pull Chain		1

Cut 45° angles on the ends, weld them from the outside, and smooth the welds with an angle grinder.

Cut the center brace (b) to length and weld it into place, then weld a gusset (e) over each of the front corners to complete the frame.

Weld the two back legs (c) and the front leg (d) into place. The front leg needs a brace (f) to give it strength. It's welded on a diagonal from the bottom of the leg up to the center brace. Thread the two nuts (g) onto the leveler rod (h), adjust the nuts to line up with the brace and the leg, and then weld the nuts into place.

Weld the axle (j) approximately ½ inch up from the bottom of each back leg. Add a washer (k) to each end of the axle, slide on the wheels, then use the nuts (l) to hold them in place.

Two carriage bolts (m), with the heads protruding ½ inch from opposite sides of the frame, are welded on the back of the frame. The burner mount wire (q) will be anchored to these.

Lay the expanded metal mesh (o) into the frame and tack-weld it into place. The bricks must lay perfectly flat in the tray. Smooth out raised areas with an angle grinder. Set bricks into the frame and set the kiln in place.

Tools and Materials

Tool List

- Medium- to heavy-weight leather gloves
- Dust mask
- Goggles or safety glasses
- Wire cutters
- Channel lock pliers
- Needle nose pliers
- Hack saw
- 18- or 24-inch carpenter's square
- Spirit (bubble) level
- Yardstick
- Permanent marker (medium point)
- Retractable-blade cutter and extra blades
- ¼- and ½-inch round files

Kiln Materials List

Burner, hose and fittings [see note 1]
Propane tank [see note 2]
7 feet of 24-inch wide farm fencing with 1×2-inch openings [see note 3]
8 J-clips (require J-clip pliers) [see note 4]
7 feet of 24-inch-wide Kaowool Cerablanket (1-inch thick) [see note 5]
13 K24 insulating fire bricks [see note 6]
28 feet of 16-gauge high-temperature nickel-chrome wire [see note 7]
2 ft. of 24- or 26-gauge steel wire [see note 8]
40 fired clay "buttons" made from 10-12 lbs. of stoneware clay [see note 9]
1/8×6×6-inch Pyrex glass [see note 10]
3 6-inch kiln posts
15-inch kiln shelf (1x15x15)

Notes

1. We used a weed burner for this kiln, but a commercial raku burner equipped with a regulator is best. The burner requires an adjustable valve.
2. Any fairly new propane tank serves the purpose. The small, BBQ-size tanks for outdoor grills tend to freeze up as they empty, so consider a larger tank or several tanks connected with a manifold.
3. Farm fencing (not chicken wire) is available at farm supply stores and home centers. It's easy to cut with a good, sharp pair of wire cutters.
4. J-clips and J-clip pliers are great for joining the ends of the farm fencing. Trouble is, you have to buy more than you need for just one kiln and the J-clip pliers are almost essential in order to use the clips. As an option you can join the farm fencing with short lengths of light-gauge wire.
5. Kaowool Cerablanket is available in 1- and 2-inch thicknesses, but the 1-inch is fine for this kiln. Purchase from a ceramic supply store.
6. Insulating fire bricks (ISB) are available at ceramic supply stores. Hard bricks are not suitable for this application.
7. High-temp, 16-gauge wire is one of the most expensive components of this kiln. Check your local ceramics supply store.
8. This is ordinary steel wire that's easy to bend and twist. It's used to hold the Pyrex glass window in place. Some ceramic supply stores stock an item called "stem wire," which is used in glass ornaments, etc.
9. For fired buttons, use cone 6-10 stoneware clay, roll it to ¼-inch thick, and cut 3-inch squares. When leather hard, drill two ⅜–inch holes a half inch apart. Use a damp sponge to smooth the edges of the holes on both sides. Use 24 buttons for the kiln frame, twelve for the lid and keep four spares. Make two half-size buttons but with the same hole placement. One goes under the burner port, and the other is a spare.
10. Purchase the Pyrex glass window from a local glass supplier/installer. We bought three for about $9 each.

Burner Mount

We used a weed burner for our kiln and devised a method to attach it. (Devise another system for the type of burner you choose.) Now for the trickiest part of the whole cart: the burner mount wire (q) and the burner mount trough (r). File a notch in one end of the trough to fit over the axle. We clamped the trough in a vise and used a ½-inch round file to cut this. Note: The notch you cut should be deep enough to hold the trough on the axle, yet open enough that the trough can be removed easily.

The trough had to be bent so that it would hold the burner horizontally in front of the burner port. We put the trough's notch back on the axle and then looked at the trough from the side to judge the height and angle of the bend. After marking the location of the bend, we put the trough back in the vise and made the initial bend. After a few trials we had it where we wanted it.

The burner mount wire (b) is 31 inches long. We marked the midpoint at 15½ inches and made a soft bend, a slight bend 2 inches in from each end, then, using needle nose pliers, formed hooks on the ends. Take your time here. The hooks have to be open enough to fit easily over the carriage bolts (m).

Hook the burner mount wire over the carriage bolts and let the mount wire rest on the floor. Next, put the notched end of the burner trough on the center of the axle, and then, while holding the trough horizontally, the mount wire was lifted until it touched the bottom of the trough. That contact point was marked so a notch, ½ inch deep, could be cut using a small round file. With the mount wire in this notch, the trough is held in place.

The burner was laid in the trough, positioned so the opening was just outside the burner port and clamped into place. With that, the burner hose was connected to the tank, we used soapy water around the connections to check for any leaks and, at last, we were ready for raku.

Using the Kiln

Before using the kiln, level it by adjusting the threaded leveler rod (h). After the first firing, we wired a couple of metal handles to each side of the kiln so it would be easy to move. We bought inexpensive gate handles and held them in place with the light-gauge wire. We also realized that it would be nice to be able to move the kiln out of the way while the kiln was still very hot so we attached a 4-foot pull chain (u) using a couple of ³⁄₁₆ inch quick links (v). Place three 6-inch kiln posts in a triangular layout to support the kiln shelf. Locate one post directly opposite the burner at the back and widely space the other two on either side of the burner.

Portable Gas Raku Kiln

by George Juliano with Elizabeth Hudgins

A portable gas kiln was difficult to find within our budget, but the school needed one if my students and I were going to experiment with raku firing. I decided that the kiln had to be simple to operate, light, easily movable, safe and effective. After doing some research and shopping, I determined that I could build one that would meet all of my criteria for a lot less than I could possibly purchase one. So, with the help of the auto mechanics teacher at my school, I set out to build the kiln.

Process

Take a 55-gallon oil drum and clean it thoroughly with Varsol. Once it's completely clean and dry, use a reciprocating saw with a metal cutting blade, and cut the drum just above the second rib (about 24 inches from the base to the rim). Cutting above the rib provides extra strength and stability to the rim (figure 1).

Next, use the saw to cut a burner port 3 inches from the bottom, making a hole $3^1/_2$ inches square (figure 2). This hole is where you'll direct the burner when you fire the kiln. Finally, cut a 4-inch-square hole in the center of the lid and drill four ½-inch holes 2 inches from the center of each side of the square.

Bolt 3 metal "spider legs" to the drum lid, using 2 bolts for each leg (figure 3). Make legs from 2-inch-wide by ¼-inch-thick strips of metal, and bend them over the edge of the lid so that 3 inches remain on the top while 8 inches of metal hang over the edge. Set the metal strips at a slight

CAUTION

File all cut metal edges smooth to avoid injury.

CAUTION

Fiber from fiber blankets can irritate the eyes, nose, throat, lungs and skin during installation and use. Always wear protective clothing and gloves, and use a respirator according to manufacturers' recommendations.

TIP

In addition to the legs, bolt 2 handles onto the lid about 3 inches from the edge so you can easily grab and lift the lid to remove raku pieces. Also, bolt 2 handles on the sides of the drum at a comfortable height for picking up and moving the kiln. This is important for taking the kiln inside between uses.

angle away from the vertical sides of the drum when you place the lid on the drum. The legs add some stability for the lid when it's in place, but they shouldn't get in the way of putting the lid on or taking it off. The major purpose of the legs is to hold the extremely hot lid off the ground when removing pots from the kiln. The legs allow air to flow under the lid, preventing it from destroying blacktop or grass.

To insulate the inside of the kiln, use 1-inch-thick fiber-insulating blanket, which is carried by most pottery supply stores (figure 4). Cut 2 circles, 1 to fit the lid and 1 to fit the bottom of the drum. Also cut a large rectangle so that the length measures the same as the circumference of the drum, and the width equals the height of the drum. Carefully place the fiber into the drum to make sure that it fits snugly with no gaps.

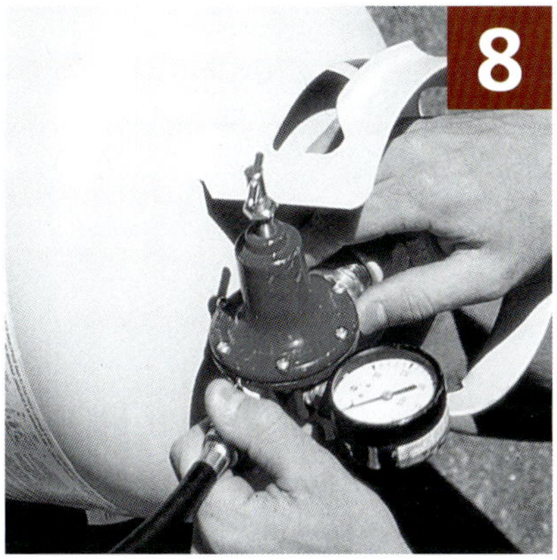

Be careful not to press on the fiber since compression of the fibers diminish the effectiveness of the insulation. Once you're certain of a snug fit, remove the fiber-blanket pieces and coat the inside of the drum and the underside of the lid with sodium silicate. Let that set for a couple of minutes, then gently push the fiber back in, again avoiding pressure so that the blanket remains fluffy. After the insulation is in place, carefully cut the fiber blanket out for the burner port and put the piece aside for later.

Cut small circles out of the fiber-blanket scrap you saved from the burner port hole. Place the circles over the bolt heads on the inside of the lid, using sodium silicate (figure 5). Set the finished kiln on three concrete blocks, one of which is located directly under the burner port. The blocks provide stability, keep the kiln off the ground and allow for a greater flow of air once the kiln is firing.

For the lid hole, carefully cut the fiber at an angle from corner to corner so that you can fold the triangles out and over the sides of the square (figure 6). This enables you to cover the edge of the metal on the square so that there is insulation all around the hole. The insulation keeps the lid temperature down a little. Insert bolts through the lid from the inside through the four holes drilled earlier and through the triangular tabs. The bolts make it less likely that the fiber blanket on the lid will respond to gravity and fall off after a number of uses.

Take three shelf posts and a round shelf to make a platform for the pieces to be fired (figure 7). Place the three posts so that one of them is angled (with a corner facing the center of the burner port). Angling the post

NOTE

To provide the smoothest surface on the inside of the kiln, install all bolts with the head on the inside so the thread and nut are on the outside.

disperses the flame to both sides for more even heat distribution. Place the shelf just above the burner port.

This portable kiln can be fired with propane or natural gas. I fire my kiln with propane, and I purchased a burner system. In addition, I picked up a 10-foot-long hose, a fitting for the connection between the hose and a propane tank, a propane tank such as those used for gas grills, a tub to put the propane tank in, a valve cock, and a pressure regulator (figure 8). Attach the pressure regulator to the hose and then the hose to the fitting for the propane tank.

Attach the other end of the hose to the valve cock and then to the burner (figure 9). Place another cement block in front of the one you placed under the burner port. Use this as the base for the burner.

To stabilize the burner, I cut a common building brick (with two holes in it) so that the brick could stand on one end, providing a nice U-shaped channel to lay the burner in (figure 10).

To add even more stability, brace 2 bricks against the burner to create an A-frame "tent" for it (figure 11). (I've been glad I did this. Several times someone has walked by and accidentally hit the hose, but the burner wasn't dislodged, thanks to the channel and the A-frame security.) Place the burner head 1 inch from the burner port. This allows for oxygen to enter the kiln with the flame, making for a hotter fire in the kiln.

Finally, remove the top and bottom of an empty coffee can and place it on top of the lid to serve as a chimney for the kiln (figure 12).

TIP

Since the propane tank tends to ice up when the kiln is firing, place the tank in a tub of water. This seems to keep the icing down a bit, making the flow of gas smoother and the firing more consistent.

Raku Glazes

by Harold J. McWhinnie

How can one have new thoughts on a process that is at least 250 years old? The real question is how can an artist working today in this method combine traditional values with a new direction?

I have worked extensively in raku over the years and what initially attracted me to the process was its apparent spontaneity, but it was the combination of spontaneity with deliberateness that made this way of working perfect for me.

Raku glazes are all relatively simple formulations, and success is more likely when working with a few recipes that are similar in composition:

Recipes

Raku Base Glaze 1

Borax	9.1%
Gerstley Borate	36.4
Nepheline Syenite	45.4
Silica	9.1
	100.0%

Raku Base Glaze 2

Borax	9.1%
Ferro Frit 3134	45.4
Nepheline Syenite	45.5
	100.0%

Raku Base Glaze 3

Borax	9.1%
Gerstley Borate	36.4
Ferro Frit 3134	45.4
Silica	9.1
	100.0%

A word of caution: Glaze recipes with large amounts of Gerstley borate, colemanite or borax (around 50% or more) may be physically unstable, as boron by itself is a glass former without other elements present. This boron glass may look shiny and smooth, it may be quite beautiful, but at the same time be unacceptably soft, abradable and unstable for functional purposes. Without enough alumina/silica to provide sufficient stability, the glaze may break down easily in mild acids, releasing whatever else is in the batch to leach into any food substance to which it is exposed. For this reason, and because of the many variables of raku firing, all raku should be considered nonfunctional or decorative unless it is proven otherwise.

Color variations of these three basic recipes may be achieved by adding small amounts of copper, cobalt, tin or iron oxide, depending upon the color desired. Copper and tin mixed in a glaze which will be reduced heavily will result in copper reds not unlike those associated with high firing.

I have been interested in raku matt effects and in the contrast point that is possible between matt and gloss, a counterpoint of effects not unlike the baroque fugue in music. To that end, I have been experimenting with raku wood ash glazes. Adding ash to raku glazes would seem, at first glance, to be almost a contradiction in terms because the addition of a quantity of ash to a glaze (over 30%) will significantly increase the melting point of the glaze itself, so that such glazes are effectively placed beyond the raku temperature range.

In order to be able to use ash/raku glazes, I have experimented with firing my work in two stages: Stage one involves adding 30% ash to the raku base glazes. The result is a very dry surface (immature glaze) when fired to 1500°F. At stage two a second or a third glaze is applied over these ash glazes and the piece is refired at least once more.

The results have been quite striking: Some very metallic matt glazes are produced; a dry ash-like residue can be perceived on the surface; and the gloss overglazes give the entire piece the look of more traditional raku work. The ash residue approaches some of the qualities of wood-fired ware, especially from high-fired anagamas. Through this method of glaze layering, the effect of wood firing can be achieved at a much lower temperature and with a lot less preparation time and physical effort.

Another raku variation I like involves rubbing a bisqued-fired piece with iron oxide, then glazing only parts of the surface. When the piece is taken from the kiln and put into a bucket filled with straw, leaves, or grass, the result is a reduced stoneware kind of surface. This use of the iron does not permit the intense black reducing effects associated with American raku pieces.

Almost any low-fire glaze (cone 06 to 04) can be made into a raku glaze by lowering its melting point. In fact, some very dull cone 04 oxidation glazes can be transformed into quite nice raku glazes by this procedure. The recipes here were developed to mature at about the same temperature (determined by eye, as is normal in raku):

Recipes

Hal's Raku Base Glaze

Gerstley Borate	60 %
Cornwall Stone	19
Nepheline Syenite	19
Bentonite	2
	100 %
Add: Zircopax	8 %

Clear Crackle Raku Glaze

Gerstley Borate	68 %
Nepheline Syenite	8
Ball Clay	8
EPK Kaolin	16
	100 %
Add: Zircopax	16 %

Red Bronze Raku Glaze

Gerstley Borate	50 %
Ferro Frit 3134	50
	100 %
Add: Tin Oxide	5 %
Copper Oxide	4 %

Dark Blue Raku Glaze

Gerstley Borate	60 parts
Ferro Frit 3134	30
Barnard Clay	9
	99 parts
Add: Cobalt Oxide	1 part

Medium Blue Raku Glaze

Borax	10 parts
Gerstley Borate	70
Ball Clay	8
	88 parts
Add: Zircopax	10 parts
Cobalt Oxide	2 parts

Light Blue Raku Glaze

Gerstley Borate	60 parts
Ferro Frit 3134	30
	90 parts
Add: Tin Oxide	5 parts
Cobalt Oxide	1 part
Rutile	4 parts

Tan Raku Glaze

Gerstley Borate	60 parts
Ferro Frit 3134	30
	90 parts
Add: Tin Oxide	2 parts
Copper Carbonate	1 part
Rutile	7 parts

Recipes

Gloss Black Raku Glaze

Borax	10 parts
Gerstley Borate	40
Soda Ash	20
Nepheline Syenite	10
Barnard Clay	20
	100 parts
Add: Cobalt Carbonate	4 parts
Copper Carbonate	2 parts

A deep blue-black which does well in heavy reduction.

Copper Blue Raku Glaze

Borax	10 parts
Ferro Frit 3134	40
Nepheline Syenite	40
Soda Feldspar	14
	104 parts
Add: Cobalt Carbonate	3 parts
Rutile	10 parts

A transparent Egyptian blue; flows well. Needs moderate to heavy reduction.

White Raku Glaze

Barium Carbonate	10 parts
Borax	20
Gerstley Borate	48
Nepheline Syenite	25
	103 parts
Add: Tin Oxide	7 parts
Zircopax	10 parts

A medium gloss with a light crackle in moderate to heavy reduction.

Green Raku Glaze

Gerstley Borate	63 parts
Ferro Frit 3134	30
	93 parts
Add: Zircopax	5 parts
Red Copper Oxide	2 parts
Rutile	3 parts

Lithium Blue Raku Glaze

Gerstley Borate	50 parts
Lithium Carbonate	21
Nepheline Syenite	24
	95 parts
Add: Tin Oxide	3 parts
Copper Oxide	2 parts

A medium blue with crackle; moderate copper luster effects with heavy reduction.

In testing, moderate to heavy coats of these glazes were painted or poured onto various pieces. Firing was accomplished in a small gas kiln, typical for raku; the work was removed for reduction (in leaves and grass) when all or most of the glazes appeared to have matured.

Designing a Matt Red Raku Glaze

by Tom Buck

By chance in the early 1990's I met Gabrielle Fisher, a raku artist from New York (Long Island). She was a guest at Mohawk College in Hamilton, Ontario. She fired in the college's tophat kiln several large bas-relief works, reducing them in straw and sawdust with pleasing results. During a coffee break, we talked about glazes. She said she needed a matt raku red, firing to about cone 08 (cone 010 rather than cone 06).

After some checking and figuring, I targeted the following Seger/Flux Unity Formula to fulfill her requirements, while allowing the use of typical raku glaze materials:

CaO	0.15	Al_2O_3	0.12-0.15	SiO_2	0.5-0.9
Li_2O	0.25	B_2O_3	1.50	TiO_2	0.15
Na_2O	0.60				

For low-fire glazes, sodium oxide (Na_2O) serves as the main flux oxide ("melter"), so that the Seger formula provides a good guide if applied with caution, since in practice the choice of materials is rather limited to borate frits or soluble materials like borax, boric acid, pearl ash, soda ash, etc.

In the formula above, unusually high amounts of alkali oxides (Li_2O, Na_2O) and boric oxide (B_2O_3) are specified; this is often demanded in raku glazes. As a result, borax (hydrated sodium borate) is usually chosen as the main ingredient. The second major flux, lithium oxide (Li_2O), may come from spodumene and/or lithium carbonate. It is also best to keep low as possible the amount of alkaline earth oxides (calcium & magnesium oxides). Oxides of tin and copper should produce red on reduction.

Matting can be achieved by adding between 5% and 10% zirconium spinel (a colorless material). Calculation yielded the recipe, Matt Red Raku Glaze.

Recipe

Matt Red Raku Glaze

Cone 08

Borax $10H_2O$	70.5%
Spodumene	17.0
Ferro Frit 3134	8.0
Lithium Carbonate	1.5
Titanium Dioxide	3.0
	100.0%
Add: Tin Oxide	5.5%
Zirconium Spinel	5.5%
Copper Carbonate	1.7%

After firing, this combination of glassmaking components (excluding additives) has the following glaze composition:

Oxide	Unity	Weight %
CaO	0.10	2.5
Li_2O	0.20	2.7
Na_2O	0.70	19.0
B_2O_3	1.38	41.9
Al_2O_3	0.16	7.3
TiO_2	0.13	4.6
SiO_2	0.83	21.9

In this recipe, borax decahydrate ($Na_2B_4O_7 \cdot 10H_2O$) has been specified; this is common washing-machine borax, which gains and loses moisture on standing unless well sealed. However, one could use the more-stable pentahydrate. a crystalline material produced for farm weed control. Any colorless spinel frit can be used; one is "zirconium spinel" a synthetic made from the oxides of zirconium, silicon, zinc and aluminum. Because of its high melting point (3100°F) this powder acts as an inert filler and matting agent, and its level can be adjusted to achieve the best matt surface without loss of color.

These glaze ingredients, all fine powders, should be dispersed in a 3-to-2 mix (methanol-to-water by volume) since this liquid mixture keeps borax from going into solution (dissolving). Because of the high wood-alcohol content, smoking and flame (including pilot lights) should be prohibited near the glaze batch. Also, when stored, it should be well covered. The glaze slurry should be on the thin side, like most raku glazes. If glycerine is available, add 10% by volume to the alcohol/water blend and thus lower the flame hazard.

In one test, a bisqued tile (cone 10 porcelain body) was dipped in the test glaze then overlapped slightly with a white crackle (80% borate frit/mineral, 10% clay and 10% Zircopax). The propane-fired kiln went to cone 06 or a bit higher, then the tile was reduced in sawdust in a covered metal pail. The result was a mottled red-bronze glaze with a satiny surface. Where the white crackle glaze overlapped the red matt glaze, the combination produced metallic copper of various hues, including an unexpected gold that merits further study (although the gold may have come from bismuth-oxide vapors given off by other pots in the kiln.)

If borax pentahydrate and a substitute for Gerstley borate are available, one could adjust the recipe as in Matt Red Raku Glaze 2.

Recipe

Matt Red Raku Glaze 2

Cone 08

Borax $5H_2O$	61.0%
Spodumene	19.0
Boraq 1	14.0
Lithium Carbonate	2.5
Titanium Dioxide	3.5
	100.0%
Add: Tin Oxide	5.5%
Zirconium Spinel	5.5%
Copper Carbonate	1.7%

I have chosen Boraq1 for its low CaO/MgO content (other GB subs usually have more). This version contains less silica and more lithia, and so the glaze may be too fluid at cone 08; if so, add 2% flint/silica and test again.

It's hard to control the actual red tint generated by postfiring reduction, since chance plays such a strong role in the reduction process. However, one might obtain a brighter red by using more copper and tin materials, say up to 3% copper carbonate and 7% tin oxide, and concurrently switching from borax to a boric acid/pearl ash mix. (That means the alcohol-water mix no longer works since boric acid is soluble in alcohol. So fresh-mix the glaze in water for immediate use only; otherwise, large crystals of pearl ash will form on standing.)

For those wishing to try it, the boric acid/pearl ash mix is shown in Matt Red Raku Glaze 3.

Recipe

Matt Red Raku Glaze 3

Cone 08

Boric Acid	34.5%
Pearl Ash	34.5
Spodumene	12.0
Ferro Frit 3134	12.0
Lithium Carbonate	3.5
Titanium Dioxide	3.5
	100.0%
Add: Tin Oxide	7.0%
Zirconium Spinel	6.0%
Copper Carbonate	3.0%

This version will likely be more fluid (watch for runs, use thin coat) and it may crackle; however, it may also give a more color and a more matt surface. Also, instead of Ferro frit 3134, one can use Ferro frit 3195 or a Gerstley borate substitute. This mix with solubles may need a "tighter" biscuit, that is, one fired to a higher cone.

Valdez Flashfiring

by John Ramer Sherrill

Valdez flashfired vase, 5 inches in height, wheel-thrown stoneware, with copper glaze, by John Ramer Sherrill. "Eyespot" patterns created by moving the torch flame in then out.

Years ago, I read a well-intentioned warning about reduction firing: "Be very careful that you do not over reduce. The novice loves to see flames and smoke belching from the kiln, but when it comes time to open the kiln, he finds blackened, ruined pots inside."

I dutifully tried to follow that advice, but fortunately had some firing disasters along the way. I say "fortunately" because to succeed at Valdez flashfiring, you've got to ruin some pots. After they're properly "ruined," you can quickly and easily transform them into vessels of vibrant color, with iridescent effects that can be achieved in no other way.

Valdez flashfiring was invented through a series of accidents. The method is rather bizarre and some traditionalists are going to be horrified; purists may faint outright.

I was throwing at the wheel while bringing a reduction firing up to temperature. Engrossed in my work, I forgot about the kiln for some time. When this awful realization finally jarred my consciousness, I dashed outside to find my raku kiln in runaway mode, billowing smoke and flame, about 200°F over temperature. I quickly shut the burners down, waited for everything to cool, then winched the top up to reveal a group of blackened pots, firmly glued to their shelves by the glazes pooled at their bases.

One blackened vase stood out from the others because there was no glaze puddle at its base. I had coated it with a recipe given to me by Robert Milner, which he called "Copper Lust." Not only was the glaze not puddled, it had not even run, though it was designated a cone 06 glaze like the others.

Examining the vase carefully, I noted a small crack, which thwarted the idea of refiring it, so I decided to subject it to an experiment I had long considered. Firing up my hand-held propane torch, I turned the flame on the blackened vase, and was transfixed as startling patterns of purple, blue and green bloomed before my eyes—the flashfiring technique had been born. Because the iridescent colors look much like oil on water and because I had fired the vase shortly after the Exxon Valdez disaster, a friend jokingly christened it the "Valdez Vase" and, naturally, the name stuck.

I was enthused by my discovery, and immediately tried the propane torch on a variety of pots, without achieving any color changes at all. I even persisted on a couple, until I actually turned some areas red-hot; all I succeeded in doing was cracking the vessels. Eventually I realized that this technique absolutely will not work unless the pot is primed first by saturating it with carbon.

I have experimented with this firing method for over a year now, wasting a lot of propane in the process. Because it is a bit tricky, I will describe what I do as well as I can, so that anyone wishing to try it will not have to duplicate my mistakes.

Firing

The objective of the reduction firing is to blacken the pots, so that in appearance they are a dark, dull, uniform, charcoal gray. If the firing has succeeded, the glaze will not be glossy at this stage.

The upper section of my kiln, a commercially available design, is winched up. The top and bottom sections fit snugly together, so that oxygen can get in only through the port at the top and from around the burners. Valdez flashfiring may be harder to accomplish with a kiln of looser construction. I fire my preferred glaze well beyond cone 06, or at least 1900°F as quickly as I can, then adjust the kiln to absolutely maximum reduction, with the most inefficient flame possible. I cover the port almost entirely, leaving an open slit of ?? inch or less through which to view the kiln interior. I then adjust the gas regulator valve so that a lick of orange flame escapes from the slit. The size of the flame is not important, but it must be billowing black smoke, as dense as you can make it. The interior of the kiln must be so smoky that your view is obscured. If burner output is visible, adjust it for long, wavering, yellow flame.

Under these conditions, the kiln will begin to cool rapidly and you may need to monitor it closely to keep reduction conditions as constant as possible until the temperature has fallen below 1200°F. Then you can shut the kiln down. Make no judgments about the failure or success of the firing until the temperature has fallen below 900°F, because even the faintest glow will fool you into thinking the pots have not blackened.

If, at this time, the pots are grass green, your atmosphere was, of course, hopelessly oxidizing. If the color is burgundy or copper, the kiln was only conventionally reduced; the glaze results can be quite pretty, but they are not Valdez. The kiln must be choked further. If the color is blue or purple, you are very close. If the final color is a uniformly dull charcoal gray, you have succeeded, and you are ready for the next stage. Call some friends over. Seeing their mouths drop open while you color your pots with a torch is great fun.

If the pots are the wrong color, you can immediately bring the kiln back up to temperature and try again. The best Valdez vase I ever made was fired six times in rapid succession the same day. I was ready to give up when it finally came out right.

Remove Carbon

When a vase turns out properly "charred," do not remove it with your bare hands; use clean gloves or a cloth, as oils from your skin will mar the glaze surface before you finish it with the propane torch.

Set the pot to be torched on a non-combustible surface. Adjust the torch flame to maximum and play it gently across the glaze surface, keeping the tip 3 to 4 inches away from the pot. Don't hold the flame on any one area, as the pot will overheat at that point and crack. Fan the flame back and forth until a soft, golden color appears. Now the carbon saturating the glaze is invisibly linking with free oxygen, thereby becoming carbon dioxide. It is best to continue this procedure until the entire pot has been freed of carbon. If some areas of bright colors appear, don't worry about them, but try to avoid it. At this stage you only want to drive the carbon out. Just remember to use a light touch with the propane torch.

A pot is properly "charred" for Valdez flashfiring when uniformly charcoal gray.

Soft golden color appears when the torch is lightly fanned over the surface.

Intense colors are achieved by holding the flame closer to the pot for a few seconds.

Bring Out Color

Let the pot cool to room temperature. If you proceed while it is still hot, the colors will be much less bright. If you are impatient, put it in the freezer. Remember, though, until it is finished, handle it with a cloth.

With the pot cooled, you are ready for the final stage. Fire up the torch and sweep it quickly across any areas that have turned green, blue or purple. They are instantly erased and are replaced by the golden color. Now bring the flame tip straight toward the glazed surface, to within an inch or so, hold it for about a second, then quickly withdraw it. Intense, vibrant colors will bloom before your eyes after the flame is removed. I have done this many times but it remains a fascinating sight.

There are many ways of manipulating the flame, and an infinite variety of patterns is possible. If you are unhappy with some area of the pattern you have created, simply sweep the flame over it (while it is cool) to erase it, then redecorate. Bear in mind, however, that each erasure means that the next pattern will be slightly less vibrant. Once you are satisfied with the patterns, you can handle the pot with your bare hands. Your fingerprints will still have some adverse effects if you later redecorate, but not nearly as much as they would have had you touched the pot at an earlier stage.

The color blooms are on the outer surface of the glaze only, and I don't yet know how they will hold up over time. I still have that vase from the first firing and it has retained its color. To ensure that ambient oxidation would not occur, I have tried coating test pieces with acrylic wax or a commercial protective spray. Both substances instantly and unacceptably muted the colors.

Not all copper glazes can be treated in this manner, but a lot of them can. The three listed below are raku glazes, but I am reasonably sure that stoneware reduction glazes would also work.

Recipes

Glaze 1

Cone 010-06, reduction

Gerstley Borate (Colemanite)	82.0%
Cornwall Stone	18.0
	100.0%
Add: Cobalt Carbonate	1.1%
Copper Carbonate	2.1%
Yellow Ocher	8.0%

Glaze 2

Cone 6, reduction

Gerstley Borate (Colemanite)	50 %
Talc	30
Nepheline Syenite	20
	100 %
Add: Copper Carbonate	3 %

This one, from Robert Keyser, matures to a pleasant matt at cone 06, but I have fired it a bit higher successfully.

Glaze 3: Copper Lust

Cone 06-02, reduction

Gerstley Borate (Colemanite)	70 %
Plastic Vitrox Clay	30
	100 %
Add: Copper Carbonate	4 %
Cobalt Carbonate	2 %

Copper Lust, the most versatile of the three, is my preferred glaze. In a normal raku firing, it yields a beautiful copper gloss. With the colorants deleted, it is a superior clear raku glaze that yields deep, dark crackles. Best of all, it absolutely will not run when fired at raku temperature.

With the proportions of the base ingredients reversed (colemanite 30% and plastic vitrox 70%), Copper Lust becomes a Cone 6 stoneware glaze. In fact, the ratio of these two ingredients can be manipulated so that the glaze matures at any point in between.

Copper Lust can also be applied to greenware directly; however, this is true only for firings with a gradual cooling period, such as Valdez, definitely not for quickly cooling raku firings.

The cone 06 version can be reduction fired to cone 02 without running, and can yield combinations of colors that include green, burgundy, blue, purple, lavender or pure copper (sometimes matt surfaced if rapidly fired).

Another great feature of this glaze is due to its nonfluid nature. It can be used to save pots that crack in the bisque firing, as long as the crack is not in the lip or on the bottom. This is true for even really bad cracks. Simply mix the glaze thickly (with the colorants listed) and trowel it into the crack, then apply an extremely thick layer of glaze over the crack, inside and out. On large vessels, I have piled it on as heavily as 3/16 inch thick. Applied that thickly, it will run an inch or so, so proceed accordingly. Not only will the crack be obliterated, but the vessel, when struck, will ring like a bell. This method of mending cracks with glaze has also proved to be quite suitable for Valdez firings.

Using these glazes, you should be able to do a perfect Valdez firing at some point in the first half-dozen attempts, instead of the many dozens of trials it took me before achieving consistent results.

The clay bodies I use for Valdez flashfiring are from James Chapelle's *The Potter's Complete Book of Clay and Glazes.*

Recipes

White Throwing Body 1

Cone 2-3

Custer Feldspar	5 %
Nepheline Syenite	10
Wollastonite	5
Kaolin	45
Kentucky OM4 Ball Clay	25
Silica	10
	100 %

White Throwing Body 2

Cone 01-3

Talc	5 %
Nepheline Syenite	20
Kaolin	20
Kentucky OM4 Ball Clay	30
Silica	25
	100 %

Oh, and don't try a Valdez flashfiring on a really windy day; you won't be able to get enough smoky reduction because there will be just too much air moving.

Interference Colors with Copper Raku

by R.W. Burrows

Some raku potters obtain striking smoke or crackle patterns, while others use copper carbonate to produce metallic effects in brilliant hues of magenta, violet, purple, blue and yellow-green. However, no copper technique has produced consistent successes. In fact, the state of this art is a morass of superstition. One potter wrote that if you want purple, use banana peels, as they contain potassium. Obviously, at one time he used banana peels and happened to get purple, but—I can assure you—there was absolutely no connection.

In "Valdez Flashfiring," J.R. Sherrill describes a process where the glaze is reduced in a fuel-rich gas kiln, then reheated with a propane torch. As many as six attempts were sometimes required to achieve good results. But my sculptures may have a hundred parts, and I do not have the energy for 600 attempts. I concluded that consistent results could only be obtained if the important variables in the process were identified, and methods of controlling them developed. To this end, a detailed investigation was conducted at Red Rocks Community College in Lakewood, Colorado.

First, a literature search revealed that the colors are the result of an optical phenomenon known as light interference. Colors are produced when the glaze is first reduced to pure copper, then reoxidized to form a thin layer of cupric oxide, about 0.1 micron (a micron is one millionth of a meter) thick.

The brilliant colors of a peacock's tail, soap bubbles and oil films are all due to light interference. To illustrate: suppose two motorboats pass in opposite directions setting up waves that cross one another. When the top of one wave meets the top of another wave, a higher wave is produced. But when the crest of one crosses the trough of another, they cancel each other out. (And there are variations between these extremes.) If the two waves add, this is called "constructive" interference. If they cancel, it is called "destructive" interference.

In the case of light waves, which are not 10 feet long, but are less than one micron in length, it is constructive and destructive interference that produces the brilliant colors. For example, red light has a wavelength of 0.7 micron while violet light's wavelength is 0.4 micron.

Light interference is produced when a part of the light is reflected from the top of the copper oxide film and the rest of the light is reflected from the underlying copper. If the copper oxide layer is 0.4 micron thick, both reflections will combine constructively to produce an intense violet color. Because of the law of conservation of energy, other colors will be canceled out. (The above statement is not quite true because light travels faster in copper oxide than in air, but it serves to illustrate the phenomenon of light interference.) The observed color depends on the thickness of the copper oxide film. A change in thickness of less than 0.1 micron will change the color.

Metallurgy books revealed that when copper is heated to 392°F, it begins to slowly react with the oxygen in the air to form a thin film of copper oxide. To explore this, a plate of copper was outfitted with an iron-constantan thermocouple and heated with a propane torch. As the oxide layer built up, the color was observed to change to magenta, then violet, purple, blue, yellow-green and finally grayish green. At slightly above 400°F, several minutes are required for this total progressive transition to occur, but at 700°F, it will instantly oxidize to a grayish green color. If exposed to air when its temperature is below 392°F, no oxide film will form.

It is appropriate here to describe the effect of the propane torch. The inner flame cone consists of unburned hydrogen and carbon monoxide. When applied to either copper or a copper glaze, the reducing action of these two gases immediately cleans off the copper oxide layer. But the instant the flame is removed, the copper is exposed to air, and interference colors will form.

Therefore, interference colors are produced by first reducing the glaze to pure copper, then reoxidizing the copper in a controlled way. Specifically, the reduction phase (in which oxygen-deficient flames lick over the pottery) completely reduces the copper oxide to metallic copper (the copper carbonate having been converted in the kiln to copper oxide). It is the reoxidation phase in which the access of air to the hot copper is severely limited in a way that produces colors by avoiding overoxidation.

Most potters believe that the reduction to pure copper is accomplished by the foul-smelling, carbon-filled smoke that is produced by smothering burning combustibles. We found that exposing pottery to this smoke did color crackle-glaze cracks and darken raw clay, but had no reducing effect on the copper glaze.

One evening, I held (with tongs) a sheet of copper above the flame in my fireplace until it collected black soot. When held in the orange-glowing part of the flame, it developed a black coating of copper oxide. But when held down between two logs where the flames were originating, it quickly showed the color of bright copper.

If the top surface of a reduced-copper glaze is ground away with emery paper, reduced-copper particles are still present below the surface. How is it that the reducing gas is able to penetrate a Gerstley borate glaze? Dr. Gerard P. Martins of the Colorado School of Mines believes this can only be accomplished by the hydrogen molecules, since carbon monoxide molecules are too large and sluggish to penetrate glaze as the hydrogen can. Obviously, hydrogen plays a major role, contrary to popular belief that the reduction is accomplished by carbon monoxide.

We concluded that the reduction of copper oxide to copper is accomplished by the unburned hydrogen and carbon monoxide within flames that lick over the pottery, in a manner identical to the propane flame previously discussed. Neither smoke nor carbon particles play a role in the reduction process.

To verify that, try this experiment: Place a hot pot in a tin can containing a small amount of alcohol. (Make sure the area around your experiment is noncombustible and fire safe.) Since an alcohol flame produces no smoke or carbon particles, it is transparent and you can actually observe the reduction taking place; it occurs very rapidly. Therefore, when a lid is placed on a reduction chamber to extinguish the flames, the copper is already in a reduced state and the main function of the lid is to cut off the oxygen supply and prevent the reduced copper from reoxidizing, as it will do until it has cooled below 400°F.

To reduce the glaze to pure copper, about 10 seconds of vigorous flames licking evenly over the pot are required. We used denatured alcohol for small objects—about ¼ cup or less for a gallon can. For larger objects, where the use of alcohol becomes more hazardous, we used loosely packed excelsior, sometimes sprinkling it with alcohol.

The combustible material should be arranged around the pottery so that flames touch all the surfaces. Plates and slabs should be leaned up almost vertically against some support. Good reduction has been obtained with as little as 3 seconds and as long as 15 seconds exposure to the flames; 8 to 10 seconds is nominal. After this, the flame should be snuffed out and air instantly excluded, as described next.

Control of Oxidation

A common raku technique is to place the pot in a garbage can filled with some combustible material; then, after a few seconds of flaming, a loose-fitting lid is put on. If anything really good comes of this, it is almost a miracle, because once the lid is placed over the can, a very insidious thing happens. As the internal temperature and pressure decline, air is actively sucked back into the can, usually reoxidizing copper beyond the point of good color.

This creates the next question: How much air re-entering the can is actually needed to produce the interference colors? The answer may surprise you. Let's say that the surface area of a pot is 1 square foot, and we want to produce the color violet. The wavelength of violet light is 0.4 micron. The index of refraction of cupric oxide (CuO) is 2.63. Constructive light interference producing the color violet occurs when the thickness of the copper oxide layer is $0.4 \div (2 \times 2.63)$ which equals 0.076 micron. This thin oxide layer, over an area of one square foot, weighs 0.045 grams and contains 0.009 grams of oxygen. This is equivalent to 0.0063 liters of oxygen or 0.03 liters of air. This is only ⅛ cup of air. Of course, only some fraction of the air molecules that leak back in would combine with the copper in the glaze, so it is reasonable to estimate that about one cup of air would produce the violet color. The main point is that very little air leaking back in is required to produce color. Even if a newspaper seal is used under the lid, much more air than 1 cupful will probably be sucked in. There is a mitigating factor, however; if, on top of the pot, there are papers, pine needles or wood shavings that are still smoldering, many oxygen molecules will be trapped before they reach and overoxidize the pot.

To overcome the problem of air leaking back into the can, we prepared a container and a lid with a perfect seal. The result was spectacularly disastrous. When the interior of the can cooled down, the internal pressure dropped so much that the external air pressure (14.7 pounds per square inch) crushed the entire can and destroyed the pot.

Since this unfortunate incident, we have tested four other methods, all of which have given consistent, repeatable results: 1) The Perfect Water Seal; 2) The Mud Seal; 3) A Weed Sprayer; and 4) Steaming Paper.

The Perfect Water Seal

The procedure is as follows: Place several bricks in the bottom of a bucket and add water. Punch a ring of small air holes about ¾ inch up around the bottom of a can. This will assure a vigorous flame over the surface of the pot. Pour 1/16 inch of denatured alcohol into the bottom of the can and set it on the bricks. Transfer the hot pot from the kiln to the can. After about 10 seconds of flaming, smother the flame with a tall container, like a metal wastebasket, taking care that the inverted rim of the container is below the surface of the water. The container will float, so put a weight on top. There must be enough separation between the water level and the pot to prevent water from contacting the pot as it rises up inside the container. Also, there should be enough water so the outer level of the water does not drop to the point where air is sucked in.

This method perfectly excludes all air, producing very clean copper surfaces, but is tricky to use. It is most applicable to working with small objects for experimental testing.

The three remaining methods are much easier to implement, and usually produce good colors, since very little air will leak back in.

The Mud Seal

Pour water on the ground. With a shovel, work up an inch or two of mud. Form a nest of excelsior, with or without alcohol. Place the hot pot in the nest and, after about 10 seconds of flaming, cover with an inverted garbage can. Push down to get a good mud seal around the can's rim. Leave in place until cool, or remove the can after a few minutes and control the color formation with a water sprayer.

A Weed Sprayer

Place the hot pot in a nest of excelsior, with or without alcohol. After 10 seconds, cover with a garbage can with a ⅜-inch-diameter hole in the inverted bottom. When gases appear to almost stop coming from the hole, insert a weed-sprayer nozzle (metal only) and continuously spray water into the can. The resulting outflow of steam will prevent air from entering the can. Keep spraying until the pottery is below 400°F, or remove the can after a few minutes to observe and control the color formation with direct water spray.

Steaming Paper

Place the hot pot in the bottom of a can containing excelsior or alcohol or both. After 10 seconds of flaming, cover with a lid consisting of a ¼-inch-thick wet newspaper topped by a board. Put bricks or rocks on top for weight. The heat from the pot will cause water in the newspaper to produce steam, which will then force its way out of the can, preventing the outside air from entering. With larger items, it may be necessary to also throw in several crumpled balls of wet newspaper to keep the steam coming out until the pot has cooled to under 400°F.

Postfiring Reoxidation

The resulting surface may have too much oxidation (predominantly yellow-green or green) or too little oxidation (pale copper and no color). If it is overoxidized, give the pot another coating of glaze and put it back into the kiln.

If it is underoxidized, put it in an oven set at 400°F; this may take 30 minutes to an hour. If no color is evident then, slowly increase the temperature until the desired colors are obtained. At this point, if you are still not pleased with the colors, attack the pot with a propane torch.

If colors do not form in the oven at 400°F-425°F, the glaze has been overfired in the kiln and too many copper oxide particles have dissolved in the glass. It is important to keep the glaze to the immature side.

Glaze Considerations

Interference colors appear most intense when the light source and the eye of the observer are both at the same angle to the plane of the glaze surface. When the glaze surface is tilted a few degrees, the intensity of the colors diminishes. This makes good photographs difficult to achieve unless an omnidirectional light source is used.

There are three approaches to overcoming this problem:

1. Matt and semimatt glazes widen the good color-viewing angle better than glazes with a smooth, glossy surface.
2. A grooved surface on the clay, such as that obtained by texturing the surface with burlap, canvas or screen wire, also widens the viewing angle for good color.
3. Underfiring the glaze, so that little nodules or lumps of partially melted glaze are present, produces a scintillating effect.

Thus, I tend to avoid very mature, shiny glaze surfaces. In addition to the viewing angle problem, the shiny surface diminishes colors in the same way that putting glass over a painting causes objectional light reflections.

One glaze that has produced good results is called Post-Pac-Man from Piepenburg's book *Raku Pottery*.

Recipe

Post-Pac-Man Raku Glaze

Bone Ash	1 part
Gerstley Borate	5
Nepheline Syenite	1
	7 parts
Add: Copper Carbonate	1 part
Red Iron Oxide	1/8 part

Parts listed in this glaze are by weight. Additionally, tests have indicated that the red iron oxide has no effect and so can be omitted. Also, increasing the bone ash to 1½ to 2 parts gives a very nice matt finish. I fire this glaze to 1800°F, keeping it on the immature side, which yields much better colors.

Raku Color and Opacity

by Jeff Zamek

Top, raku base glaze with 10% Mason stain 6308 Vivid Blue; middle, raku base glaze with 10% Superpax; bottom, raku base glaze plus 10% Mason stain 6271 Mint Green.

A simple approach to color development and opacity in glazes, coupled with the fast-firing raku process, can greatly aid in learning about the effects of adding metallic coloring oxides, stains and opacity-producing agents to a base glaze. Since you can easily mature glazes in a raku firing in 20–30 minutes, you don't have to wait long to find out what happens when you add copper, chrome, cobalt or any number of other coloring oxides or stains to a glaze. The pot is removed red hot and either oxidized (left in the air to cool) or reduced in a combustible material. When reduced, the burning materials pull oxygen from the metallic oxides in the glaze resulting in color variations.

The raku technique is fast and dramatic, and the speed of the process compresses the time between the cause and effect of adding and subtracting various glaze materials. In addition, the results learned from these glaze tests can be applied to other temperature ranges.

Raku Glaze Requirements

Raku glazes are appreciated for their ability to offer various colors and textural variations, however, there are several requirements for an effective glaze.

- For safety reasons, lead or lead-based frits are not recommended for raku glazes even though recipes appear in many older texts.
- Raw materials should be readily available.
- Soluble glaze materials such as Gerstley borate, colemanite, borax, pearl ash, soda ash and boric acid can be used; however, they can leach into the glaze water resulting in unpredictable glaze effects. In some instances,

the results can be aesthetically interesting, but soluble glaze materials can also cause blistering, running and dry areas in some glazes.

- The glaze should stay in suspension for a reasonable length of time to allow for glaze application.
- Raku glazes must be able to mature over a wide range of temperatures because not all kilns fire evenly or accurately.
- Raku glazes must be able to withstand variations in reduction atmospheres.

Getting Started

Just a few raw materials are needed to produce a glaze with different colors and surface textures. For example, you can make 1½ gallons of a base glaze with 8 pounds of Ferro Frit 3110 and 2 pounds of EPK kaolin. Superpax can be purchased in 3-pound increments. Some metallic coloring oxides and stains are more expensive, but they can be ordered in ¼-pound increments. For small tests, use a 300 gram batch, which is suitable for several vertical test tiles.

A base glaze formula (one that's clear, semiopaque or opaque in light transmission, and glossy, satin matt or matt in surface texture) is the foundation for any addition of metallic coloring oxides, stains or opacity producing agents. It offers a known constant to judge the effects of color and opacity. In the raku

Recipe

Clear Gloss Crackle

Ferro Frit 3110	80 %
EPK Kaolin	20
	100 %
Add: Bentonite	2 %

White Opaque Gloss

Blue Variation

Superpax	15 %
Mason stain 6306 Vivid Blue	10 %

Green Variation

Mason stain 6271 Mint Green	10 %

Ferro frit 3110 provides the flux you need in the glaze for it to melt. In fact, at the raku temperature range, frits are complete glazes by themselves.

EPK kaolin, which is a clay containing alumina and silica, prevents a molten glaze from running down vertical surfaces.

Stains provide color to the base glaze.

Superpax is an opacifier that makes the clear, gloss, transparent base glaze an opaque, gloss white.

temperature range (cone 010–cone 04 or 1657°–1945°F), any frit can be considered a glaze by itself since it contains a combination of flux oxides, alumina and silica needed to form a low-fire glaze. Additions of clay can increase opacity and matt surface texture while the Superpax further aids in opacity.

Any of the metallic oxides (iron oxide, cobalt oxide, copper oxide, chrome oxide, nickel oxide, iron chromate, manganese dioxide, rutile) or their carbonate forms (copper carbonate, cobalt carbonate, manganese carbonate or nickel carbonate) adds color to a glaze. Adding approximately ⅛ percent yields a tint of color, 5% provides color with medium intensity and 10% imparts intense color to the glaze. The advantage of using just one base glaze aside from the simplicity of understanding cause and effect when using ceramic materials is that glaze color variations can be overlapped with a good chance of compatibility. And testing glazes on vertical test tiles will provide valuable information on how runny the glaze is.

Adjusting the Base

You can adjust the glaze by adding 5, 10 or 20 parts kaolin to the base for a stiffer glaze. Conversely, deleting 5, 10 or 20 parts of kaolin from the base makes the glaze more fluid. The spacing of the crackle lines (a fine network in the glaze due to tension over the clay body) depends on the fit of the clay body and glaze upon cooling. The rate of contraction varies with different frits causing craze lines in the glaze to enlarge or shrink. The rate of contraction in the clay body also affects the craze line pattern. Selecting a different frit with a different chemical composition may alter the colors obtained from metallic oxides or stains.

Raku Glaze Trials

by Lila Bakke

A technical ceramics class was a revelation to me, unlocking the mysteries of the art of pottery, from mixing clay bodies to building a gas-fired kiln. Previously, I had merely thrown pots, then handed them over to chance. Someone else had mixed the clay and glazes, and fired the kiln. But in this class, we experienced the chemistry of ceramics-not only in theory, but in the mind-wrenching, frustrating hours of research, calculations, experimentation and the rare, often unexpected moments of beauty and triumph.

One experiment was to formulate two original glaze recipes, and from them generate 20 different colors. Raku glazes captured my interest because that type of firing was something I'd never done before, and results appeared sooner. In one afternoon, I could mix and test any number of recipes; whereas at cone 10, I'd have to wait three days for the fired results.

First, I formulated a raku body that would remain white (like porcelain) after firing and would smoke from dusty gray to charred black:

Recipes

Raku Body 1

Talc	15 %
Fireclay	25
Kentucky OM4 Ball Clay	15
XX Sagger Clay	30
Silica Sand	15
	100 %

Suitable for handbuilding; 7% shrinkage at cone 08. Add 0.05% shredded nylon fibers for greater strength.

Raku Body 2

Talc	15 %
Fireclay	40
Kentucky OM4 Ball Clay	20
XX Sagger Clay	20
Silica Sand	5
	100 %

For a less abrasive and more plastic throwing body, decrease the sand and XX Sagger clay, while increasing the ball clay and fireclay content.

The next step was to find those two base glaze formulas. We had learned that calcium oxide, a common and inexpensive material, makes glazes hard and durable. Because I had never worked with it and because of its tendency to enhance certain colors, I decided it would be the predominant oxide in both glazes. But the melting point of calcium oxide is very high (2572°F), so I needed another substantial flux. For the first experimental recipe, I chose Ferro Frit 3134, mostly for its nontoxicity. For the second, I chose Kona F-4 feldspar.

The first recipe (followed here by its empirical molecular formula) also included boric oxide for further color effect:

Thinking these recipes were calculated to fire appropriately, and because I was pressed for time, curious and inexperienced, I mixed color tests before actually firing the base recipes. The colored glazes were fired on bisqued rods of Raku Clay Body 1 in a small (2-cubic-foot) gas kiln. Some were smoked; others were oxidized. Those that were smoked were covered with (and thus carbonized by) sawdust and straw.

The first test's accuracy can be disregarded, as I fired without a pyrometer or cones. But it was still valuable for experience in firing the kiln, and also because it gave indications as to the colors I'd end up with. Underfired as the glazes were, the colors were not entirely undesirable. Those based on the first glaze adhered tightly to the clay. Those based on the second did not; they were like chalk.

The second test was much more controlled because I used a pyrometer. From the first test results, I had chosen 16 colors, then added four more. Four of the bisqued rods were devoted to the two base glazes-one rod each to smoke; the other to oxidize. The rods were fired in groups

Recipes

Experimental Raku Base Glaze 1

Barium Carbonate	4.7%
Whiting	28.0
Zinc	1.9
Ferro Frit 3134	29.0
EPK Kaolin	12.4
Silica	24.0
	100.0%

0.10	KNaO	0.10	Al_2O_3	1.50	SiO_2
0.80	CaO	0.20	B_2O_3		
0.05	BaO				
0.05	ZnO				

Experimental Raku Base Glaze 2

Barium Carbonate	4.3%
Whiting	34.8
Zinc	1.8
Kona F-4 Feldspar	27.7
EPK Kaolin	19.5
Silica	11.9
	100.0%

For the second glaze, the boric oxide was omitted and the amount of alumina increased to dull color and enhance opacity:

0.10	KNaO	0.30	Al_2O_3	1.50	SiO_2
0.80	CaO				
0.05	BaO				
0.05	ZnO				

of six to eight. When the glaze appeared to be mature, I removed the rod from the kiln and recorded the temperature registering on the pyrometer. These results, though interesting, were not exactly what I had wanted; only a few matured at cone 08. Generally, Base Glaze 1 matured at cone 06-05, while Base Glaze 2 matured somewhere above cone 03.

To extend the second test, I selected nine color variations of the first recipe and ten of the second for more controlled testing. Each of the Base Glaze 1 variations was fired on four rods (for a total of 36 rods). Working with groups of four or five, I fired each of the four sets to a predetermined temperature-set one at cone 06-05; set two at cone 07; set three at cone 05; set four at cone 04. Set one was left in the kiln for ten minutes to see whether sustaining temperature for any length of time (soaking) would make a difference. It didn't. For those glazes based on the Base Glaze 2 recipe, only one rod was used for each, but none even approached maturation at 2000°F. The colors based on the first recipe were more developed, varied and far more interesting.

Regrouping my notes, I recalculated eight variations of Base Glaze 1, fluxing it down to mature at cone 08. In the first variation (A-2), the calcium oxide content was decreased, while more weight was given to the other oxides. In both A-3 and A-4, the boric oxide was increased; but in A-3 the barium oxide was eliminated, while A-4 included less silica. In A-5, the silica content was decreased, and the potassium-sodium oxide was eliminated in A-6. The last three (A-7, A-8 and A-9) were based on A-5.

Again using bisqued clay rods (three for each glaze), these recipes were fired to three different temperatures: set 1 at cone 08; set 2 at cone 06; set 3 at cone 04. A-3, A-4 and A-6 matured at cone 08: while A-5 matured at cone 07. The others matured at cone 06 or above, and were subsequently set aside.

The three that did mature at cone 08 plus A-7 (which matured at cone 07 and had an interesting texture) were retested on flat tiles. All four glazes adhered well, and each presented interesting qualities: A-3, a smooth, opaque milky white, crawled unevenly; but it left glaze in the cracks, not raw clay. A-4, also an opaque milky white, was smooth and sparkled like fresh snow. A-5 clung to the clay body, but cracked severely and looked dry. While very similar to A-3, A-6 cracked in a more regular pattern, left raw clay in the cracks and looked shinier.

I was pleased to have found four such interesting base glazes. Next on the agenda will be a series of color tests to see what effect the addition of various oxides will have on their maturation temperature.

Recipes

A-2 Raku Glaze

Barium Carbonate	9.3%
Whiting	22.8
Zinc	3.8
Ferro Frit 3134	28.4
EPK Kaolin	12.2
Silica	23.7
	100.0%

0.1	KNaO	0.1	Al_2O_3	1.5	SiO_2
0.7	CaO	0.2	B_2O_3		
0.1	BaO				
0.1	ZnO				

A-3 Raku Glaze

Colemanite	6.6%
Whiting	24.7
Zinc	3.9
Ferro Frit 3134	28.6
EPK Kaolin	12.3
Silica	23.9
	100.0%

0.1	KNaO	0.1	Al_2O_3	1.5	SiO_2
0.8	CaO	0.3	B_2O_3		
0.1	ZnO				

A-4 Raku Glaze

Barium Carbonate	4.9%
Colemanite	6.9
Whiting	26.6
Zinc	2.1
Ferro Frit 3134	30.3
EPK Kaolin	13.0
Silica	16.2
	100.0%

0.10	KNaO	0.10	Al_2O_3	1.20	SiO_2
0.80	CaO	0.30	B_2O_3		
0.05	BaO				
0.05	ZnO				

A-5 Raku Glaze

Barium Carbonate	9.5%
Colemanite	13.2
Whiting	32.1
Zinc	3.9
EPK Kaolin	12.4
Silica	28.9
	100.0%

0.10	KNaO	0.10	Al_2O_3	1.20	SiO_2
0.80	CaO	0.20	B_2O_3		
0.05	BaO				
0.05	ZnO				

A-6 Raku Glaze

Barium Carbonate	12.5%
Colemanite	11.6
Whiting	24.1
Zinc	5.2
EPK Kaolin	11.0
Silica	35.6
	100.0%

0.70	CaO	0.10	Al_2O_3	1.50	SiO_2
0.15	BaO	0.20	B_2O_3		
0.15	ZnO				

A-7 Raku Glaze

Colemanite	7.5%
Whiting	36.3
Frit 5301 (Ferro)	28.2
EPK Kaolin	5.2
Silica	22.8
	100.0%

0.2	KNaO	0.1	Al_2O_3	1.2	SiO_2
0.8	CaO	0.2	B_2O_3		

A-8 Raku Glaze

Colemanite	20.1%
Whiting	4.9
Zinc	11.9
EPK Kaolin	19.0
Silica	44.1
	100.0%

0.8	CaO	0.1	Al_2O_3	1.2	SiO_2
0.2	ZnO	0.2	B_2O_3		

A-9 Raku Glaze

Barium Carbonate	18.0%
Colemanite	12.5
Whiting	30.4
EPK Kaolin	11.8
Silica	27.3
	100.0%

0.8	CaO	0.1	Al_2O_3	1.2	SiO_2
0.2	BaO	0.2	B_2O_3		

Raku Clays

by Bill Jones

The raku firing process requires a porous nonvitrified clay that can withstand rapid heating to low-fire temperatures (cone 09–04 or approximately 1702°–1958°F) and cooling without cracking or breaking from the thermal shock. By this definition, any clay that can withstand such stresses can be considered a raku clay; however, some clays will provide a greater degree of success, especially those with a high fireclay content. It's important to recognize that when deciding on a suitable raku clay, your chances for success increase with bodies specifically formulated or adjusted for the raku process.

Additions

Most clay bodies can be used for raku by adding up to 50% grog. Grog, which is crushed, fired clay, opens the body thus making it less prone to thermal shock. Other materials you can add to increase the ability of a body to withstand thermal shock include silica sand and kyanite. Silica sand can be used as a substitute for grog, especially at the lower raku temperatures. Because it is not as absorbent as grog, you may recognize some gains in plasticity. Kyanite is a refractory material that matures at about cone 36. One of the characteristics of kyanite is that it expands when heated, which counteracts some clay shrinkage. It also creates a network of needle-like crystals allowing you to make larger forms. Commercial raku clay bodies contain one or a combination of these ingredients depending on other qualities you need in a body, such as the ability to make large work, or whether you're throwing or handbuilding. You can add any of these materials to an existing body simply by wedging them in. Using a pug mill or clay mixer provides a more homogenous mix, and dry batching provides a more consistent mix.

Bisque Firing

Most raku clays can be fired as high as cone 6–10 since they are formulated as stoneware clays. But clay is clay and it should be noted that when bisque firing for raku, you should not bisque fire higher than cone 04 (1958°F). Between approximately 2012°F and 2192°F the material becomes more dense and glasslike, thereby losing some of its ability to withstand thermal shock.

Tips for Buying Raku Clay

Most clay suppliers offer a range of raku clay bodies that can usually match the qualities you're familiar with in your regular body. They will be able to guide you either through their product descriptions or in consultation, and many clay producers will even custom blend a clay from your own recipe.

Plasticity: Clays with coarser grog are more suitable for handbuilding, while finer grog makes a better throwing clay. Some commercial clays are suitable for both.

Thermal shock: The larger and thicker the pieces, the more suitable and shock resistant the clay has to be. Increasing the shock resistance means adding more nonplastic refractory material, which may decrease plasiticity.

Color: The color of the raku clay body influences the colors of your glazes. Light-colored or buff bodies produce lighter, more brilliant glaze colors and bring out subtle shades, while darker clay bodies have a more muting affect. Raku clays made from buff stoneware, kaolins and ball clays produce lighter colors, while adding earthenware clays and colorants like iron oxide or burnt umber will create darker bodies.

Texture: Adding grog to a clay body affects the texture since the grog is already fired and does not shrink. While this is not much of a factor with fine or medium grog, it is more noticeable with coarse grog. If you're looking for smooth texture, you'll want a body with fine grog, sand or kyanite. You can also create unusual textures by wedging in sawdust or paper pulp, which also opens the body and increases resistance to thermal shock.

As with any clay or glaze, you should test a raku body before you invest a sizeable amount of energy in forming work. Many suppliers offer samples that you can test, or you can try out different additions to your existing body. Mark all of your samples and keep good records of your results.

Large-scale Raku

by Carl Gillberg

Carl Gillberg with a pair of raku jars commissioned for the Albuquerque Hilton Hotel.

I'm a potter. I make pots. Because of their large size, some people insist on calling my pots "sculpture" or "pieces." Whether or not they are art depends on the intention of and the effect on the viewer. The ones at the Sheraton Tokyo Bay are in glass cases and have bamboo or fine old Japanese garments in and on them. They are accompanied by Sam Francis and Laddie John Dill paintings. Art or decoration? Where is the line? Does it matter?

My pots are functional. Mostly they function as accents or decoration. But they look wonderful in the right context, and they make people feel good-that's nice, even if the feeling is fleeting or on the edge of the viewer's consciousness.

I've always been fascinated with the challenge of "big." (The presence of a large pot can be formidable.) About ten years ago, I made several 3-foot-diameter bowls and numerous large stoneware containers. All the shrinkage that happens at 2400°F certainly has its own integrity, but when you weigh the costs of warping and cracking against the flexibility of firing at lower temperatures, raku is the only way to go.

Making huge, heroic-scale pots (to 7 feet in height) is a rush. I begin by preparing the equipment and work area—everything I'll need for eight to ten hours is at hand.

Medium-size pots (to 5 feet high and 3 feet wide) are much more controllable. No adrenaline; just solid satisfaction knowing I'm 99% in charge. But it's still hard work. I gauge my mental and physical preparedness not only by results, but also by how many pounds of clay I can throw successfully in one day. Four to six hundred pounds is tops.

Forced drying (note steam rising from a freshly thrown base) allows Gillberg to assemble 7-foot pots in a single day

After the bottom is fired, a thrown wall section is attached and the seam trimmed inside and out.

The process is repeated for additional sections; each time the supporting wall is partly dried with a kiln burner

Gillberg uses a long, flexible board as a rib to smooth, redefine and unify the profile of previously joined sections

Hoist ropes threaded through eyebolts at the bat edge allow sections to be lifted and inverted without distortion

Each inverted section's lip fits into a U-shaped rim slot; the clay is then wire cut from the bat and a new rim is thrown.

That's enough to leave me tired the next day.

From time to time, I've played with decoration (both graphics and form distortion), but never achieved the level of satisfaction I get from tampering with the firing. Raku, the way I do it, allows me to watch the whole process-from chalk to sinter to fluid to taffy to run to flow and back to freeze. By understanding these dynamics, I am able to alter, influence and actually use the firing process as an adjunct to the other elements that make up the glaze palette.

To illustrate, the following describes how I apply a basic copper red: Combine some tin and copper, a little clay, some lithium carbonate and a good frit with water; put it all over the pot (a little thinner on the shoulders). Come back on top of that

Stair-stepped, portable scaffolding gives Gillberg variable height and firm footing when working on the mouth and shoulder of a 7-foot vase.

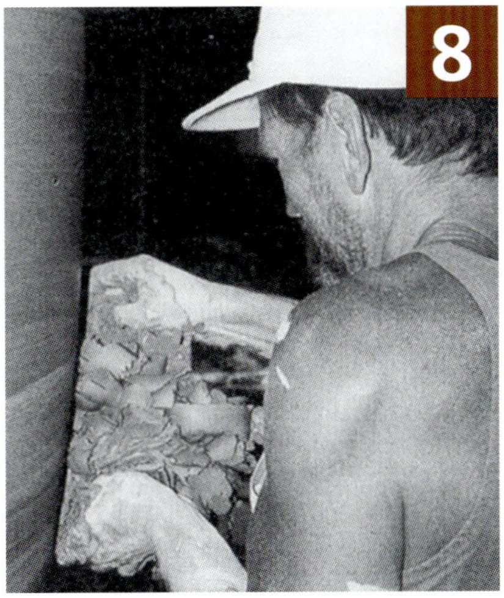

Once the basic form is established, the surface is again ribbed to finalize the profile and trim off any remaining irregularities.

Because Gillberg forces drying throughout throwing and assembly, overall moisture is equalized by covering the piece with a plastic sheet.

With the right equipment (heavy-duty electric wheel, portable scaffolding, hand-cranked and electric hoists), making "heroic-scale pots is a rush" for studio potter Carl Gillberg.

Carefully supported by a balanced system of tied ropes with cardboard padding under the knots, the green pot is hoisted from the wheel to the kiln pad for bisquing.

The fiber kiln is winched into place. During a glaze firing, the kiln top will be opened and a cardboard-lined reduction chamber (foreground) lowered over the glowing pot.

and lay the same glaze without the tin on the neck and shoulders, then back again with some copper carbonate in water (milky consistency) placed here and there. Throw a gram or two of cobalt and chrome in the same glaze without the tin, and dust the neck and shoulders. Lay down a belt of base glaze (no copper or tin) with 5%-10% opacifier, wandering off the shoulder like the neckline of a strapless gown.

Fire it until it boils. Hold the temperature there until the glaze breaks and runs. Choke the kiln until it belches orange flames for a while. Close it and shut off the gas until it cools to dull red. Crank the

At about 1800°F, the lid of the kiln is raised so that Gillberg can see if the glaze is ready for reduction.

Glaze on pots up to 4 feet high may be checked by lifting the chamber; this dramatic photo of a larger pot was a gamble.

kiln back up until the glaze is fluid again. Choke it and shut off air to the burners so that it billows thick black smoke for a while. Open it up.

You'll see transparent taffy yellow with a little light rust here and there. The neckline will have stretched down the side, and there will be lightning bolt patterns spider webbing all over like a dry lake bed. That little bit of rust will immediately start blushing, going deeper and deeper in patterns like waves or fish scales caused by the boiling, collapsing bubbles. Reduce with combustibles at the right time and you'll get just the right amount of flashing; don't reduce and the colors will be less muted, more vibrant.

Action painting? It's more like alchemy. I do it because it's fun—what one does is often less important than how it feels. What you've done matters less than what you'll do.

In Pursuit of Form

by Kristin Doner

PHOTOS: JAMES PATRICK DAWSON, MAXIMAGE

"Red and Black Amphora," from the "Imperial Pinchpot Series," 11 inches in height, with terra sigillata and reduction stencil effects.

My journey with clay has been a process of discovery and rediscovery. At times, the process has been forced, leading me to produce works that I consider overengineered, forms that stand apart from those that emerge from a more intuitive process. When I disengaged my conscious thought and allowed my work to develop through a more intuitive process, the result was a more satisfying expression, one that speaks from a deeper part of myself.

I am often surprised how intuitive responses interact with conscious inspirations to produce unexpected outcomes. For example, ancient history has been an inspiration for me. I feel connected with the past when examining an ancient artifact, realizing that the item in my hand was likely an integral part of someone's life centuries ago. I am also fascinated by the impact of time and environment on the surface of these artifacts. Frequently, organic textures and earthy colors appear in random fashion, having developed from centuries of exposure to the elements. There's little doubt that these artifacts have had an influence on my work, as witnessed by my choice of glaze treatments and my classical use of form.

The "less is more" aesthetic has also had an effect on my work. I try to reduce an expression to the pure essentials, to focus clearly on the main point of my work—form. My goal is pure simplicity of form achieved through gently sloping lines and as few distractions as possible.

To develop my sensitivity for form, I used to sit with pencil and paper sketching half silhouettes. These half forms trick the eye in a wonderful way. By building upon the visual information of one half of the form, the mind completes the other half. The result is an instant critique of the essential elements.

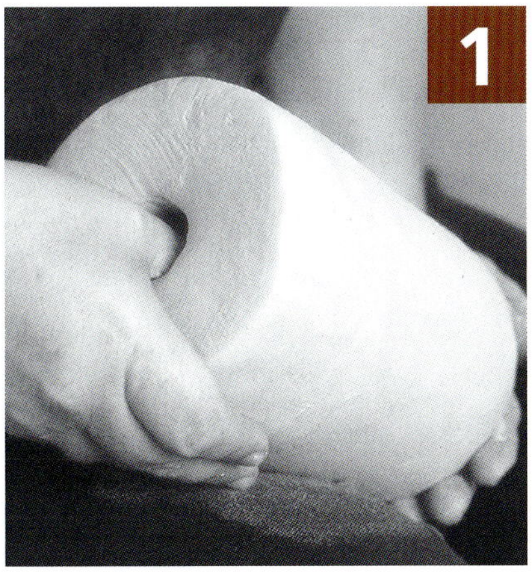

A cylindrical ball of clay is opened with the thumb. Using up to an 8-pound ball of clay for a pinchpot has become second nature.

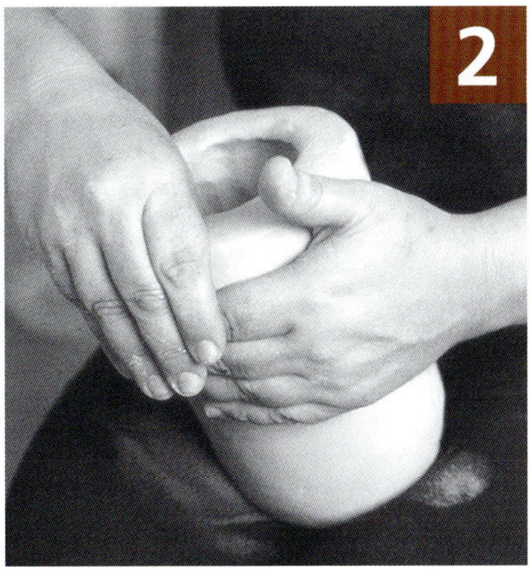

The upper walls are thinned by pushing with the thumb from the inside, while supporting the outside with the outer hand.

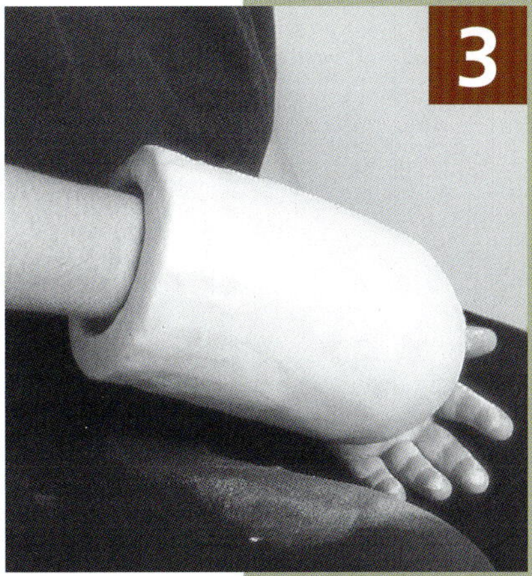

The bottom is thinned by pushing with the fingers or knuckles of the inner hand toward the outer (support) hand.

Zeroing in on an intriguing form, I used to sketch out every aspect of the completed piece in full detail, then apply the appropriate techniques to bring the expression to life. These conscious exercises were interesting to a point, but they lacked the spirit and vitality that I hungered for in my work. Eventually, I put down the pencil and started working intuitively. Letting go in this way allowed me to develop beyond the limitations of overengineered ideas.

Process

As I found myself turning repeatedly to the pinching technique, I decided to start producing these pinchpots, but on a larger-than-usual scale. I was determined to retain the simplicity of form while emphasizing its essential elements. As I allowed my intuition to take charge of developing form, I was able to focus on technique in a more active way.

When I began working with 2- or 3-pound balls for each pinchpot, the increase in mass demanded entirely new forming strategies. Use of a paddle, the size of the paddle, and the balance between drying and working times, all had an impact on the outcome.

Now, using up to an 8-pound ball of clay for a pinchpot has become second nature to me. Starting with a cylindrical ball of clay, I open with the normal pinching methods. Then I begin thinning the wall by stroking and stretching the inside, while rhythmically paddling the outside. After the clay has been allowed to dry somewhat, the same method is used to thin out the walls and develop the form, alternating between drying and forming.

The next challenge I encountered was how best to present these forms. Since the rounded bottoms are an in-

The wall thickness is evened (bottom to rim) by rhythmical paddling while slowly rotating the form with the inside hand.

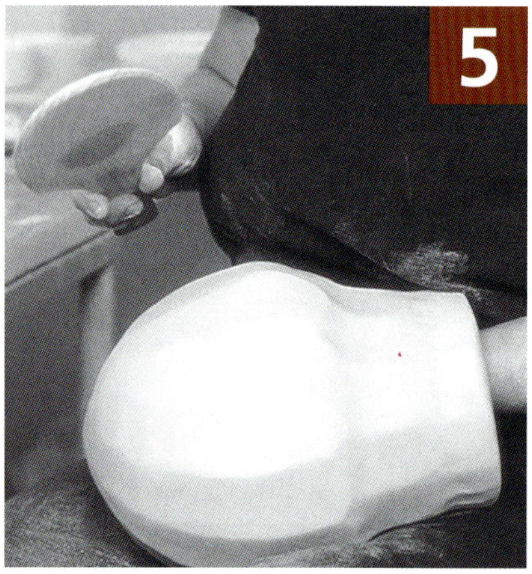

After the clay has hardened somewhat, the lower wall is stretched more, reserving a small amount of clay at the bottom.

Once the clay has firmed up again, the reserve at the bottom is stretched by stroking in the opposite direction.

tegral part of the overall statement, it is essential to incorporate a foot that doesn't detract from the form. I found that elevating the form on an understated tripod emphasized the sensual form. This also brought into play the idea of negative space subtly framing the simple roundness of the bottom.

Lugs and Lids

With the main form and presentation issues worked out, I went on to tackle the addition of handle lugs and lids. Originally, my forms were much rounder, more elliptical in shape, using handle lugs to secure an arching handle over a low-lying lid. When the vessels took on the more elongated shape, the arching handle conflicted with the more dominant form. Also, the introduction of a peaked lid mirrored the curves of the body. These new elements inspired me to lower the handle until it touched the lid, making it "work" visually, while expressly denying use.

At first, I rebelled against binding the lid to the vessel, but ultimately I was too intrigued with the result to stop doing it. Then I started feeling differently toward the vessels entirely. I began to identify with the sense of mystery this created. Upon reconsideration, I realized that binding the lid underscored the ritual aspect of the vessel as well as the concept of the vessel as an art form. However, even after I decided to tie the lid to the vessel, the possibilities for lug treatment still seemed endless. The handle lugs needed to complement the feature of the bound lid without interfering with the lines of the body. As a solution, I repeated the shape of the legs for the handle lugs, keeping the number of new elements to a minimum.

Then, the form is inverted and paddled to compress the clay and refine the contour of the bottom.

When the bottom is firm, the form is supported in a padded bowl, and paddling of the upper wall is begun.

Subsequent passes refine the curve of the shoulder; eventually, the paddle is used simply to control the amount of stretching.

Surface

Another concern in following through with form was glaze choice. After exploring the more traditional raku glazes, I settled on two basic glaze treatments: a lithium-based glaze, and reduction stenciling. While the lithium glaze is fascinating on its own and lends an ancient look to the piece, it still allows the fullness of the form to be the dominant feature. The form is even more pronounced with the use of the reduction stenciling technique. By leaving legs, lid and handle bare of glaze, I achieve a contrast that further accentuates the form. The use of terra sigillata and colored clays expands my palette, yet maintains simplicity, allowing the form to remain the focus.

Firing

The final considerations were firing issues. The tripods caused considerable difficulty to the removal of ware from a traditional raku kiln. The

The raku firing pit with two firing positions (here the left side is prepared for the next load, while the right is firing).

likelihood of jarring one of the legs while setting it down needed to be eliminated. The fact that I had to fire my work without an assistant exacerbated this problem. However, it was solved by building a removable fiber chamber with a stationary firebox in a sand pit. After the firing chamber is removed, reduction material (a newspaper log) is placed

"Unearthed Amphora," from the "Imperial Pinchpot Series," 12½ inches in height.

in the firebox. Then a large galvanized trash can is placed over the ware, and the rim sealed with sand. Making the sand pit large enough to accommodate two firing positions allows one load to be reduced while the next load is firing.

By allowing my passions to influence my work, I have developed my ceramic art in a direction that I most likely would not have intentionally chosen. It has been an exhilarating journey, one which I hope has only just begun.

Teapots and small jars, to 8 inches in height, by Kristin Doner.

Raku Reduction Stenciling

by Jerry L. Caplan

Raku platter with slip stenciled pattern.

The idea for reduction stenciling began with rust scale which I found peeling from a weathered 55-gallon drum. A few pieces were placed into a freshly glazed raku bowl with the expectation of some interesting results on firing. When the bowl was removed from reduction, big sections of the glaze fell out, leaving a network of black outlines around areas of unreduced clay. Portions of the rust and glaze clung to the bowl and could only be painstakingly removed. A more reliable stencil coating was needed-one that would fall off completely. I have since developed a slip stenciling technique which improves

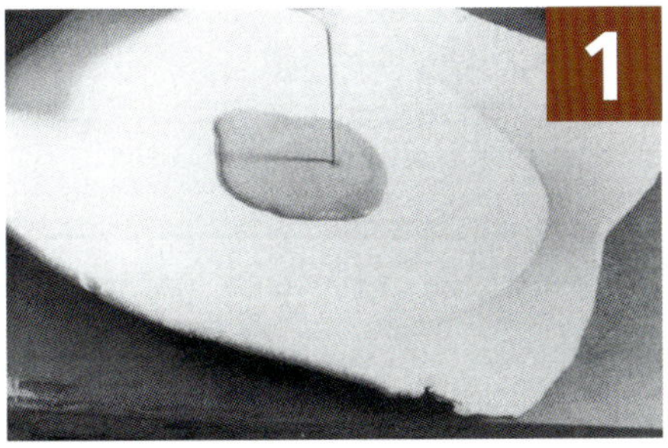

A bisqued platter is evenly coated with slip.

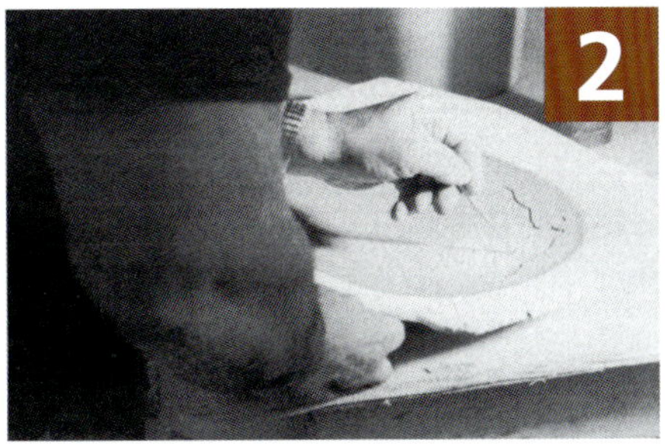

The author scratches through the damp surface using a potter's needle.

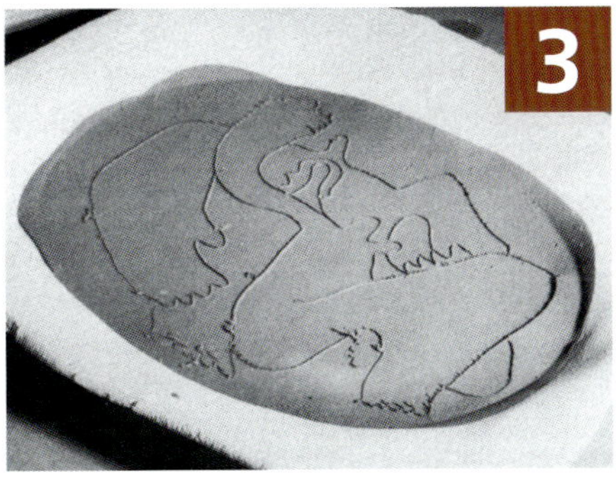

When the drawing is completed, the piece is set aside to dry rapidly

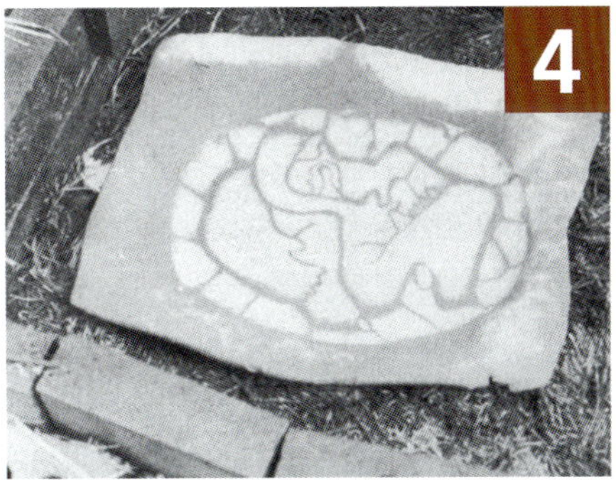

For easy access, the raku reduction pit consists of upright bricks covered with a sheet of tin.

After thorough drying, pieces are placed in the raku kiln and gently fired to temperature with the dried slip stencil in place

on the rust method and offers a variety of decorative possibilities.

A raku body is first coated with porcelain slip to give greater luminosity and accentuate contrast. Most any stoneware body with the addition of ten to fifteen per cent medium grog will suffice, and any porcelain slip may be used, provided its shrinkage is compatible. After bisque firing, the interior of the bowl is coated with a finely sieved slip made from the original raku body, and a pattern is scratched through this surface with a needle tool. As the slip dries, the solid parts of the pattern pull away from the bowl.

After thorough drying, cracked sections may be selectively removed like pieces of a jigsaw puzzle. Ware is then placed in the raku kiln at a temperature as high as about 1000°F and fired to approximately 1800°F. In the typical raku manner, objects are next removed from the kiln with tongs, and placed in a closed container filled with a combustible material (straw, grass, etc.) to produce the dark carbonized patterning.

Bowl with grid pattern, reduced porcelain slip surface on raku clay body.

The dried cracked slip can slide to the bottom of a deep bowl, giving an interesting but virtually uncontrollable effect. Deeper bowls should be handled gently while stacking and removing from the kiln, as jarring can disarrange the pattern. Flat or roughened surfaces encourage sticking of the slip, and may make it difficult to remove the clay stencil. Smooth, shallow concave surfaces seem to work best.

Many related techniques remain unexplored. A combination of reduction stenciling and glazing, wax resist under the slip coating, arranging thin shards of clay or pieces of metal inside the bowl, the use of colored slip, and many more ideas can be utilized by those wishing to advance the process through further experiments.

The completed reduced pattern, with slip stencil removed.

Ramon Camarillo
A Marraige of Three Cultures

by Cathy Grubman

"Tsunami," 18 inches in height, wheel thrown, raku fired.

Raku artist Ramon Camarillo has a lot of secrets. But don't worry. "Makamaka!" He'll tell all.

When I met Camarillo at Creative Clay Studios in Alexandria, Virginia, he was dressed in a Hawaiian shirt and beach shorts, looking less like a potter and more like a guy who was about to hang ten on a surfboard. However, when it comes to pottery, he has a lot to say, and if you are patient and welcome his spirit of "Aloha," you will learn all his secrets.

In his Hawaiian English—pronouncing raku as "raw-koo"—he happily explains how he gets that crimson red and those black-as-night areas near the base and rim of his giant pots (with "newspapers, lots of flames and a good tight lid"). Want to throw a pot as high as the entire length of your arm in one fell swoop? No problem. He'll show you how to center 25 pounds of clay without unduly stressing your back.

Laura Silberman of Potomac, Maryland, a participant in one of his kilnbuilding workshops last year, says Camarillo's philosophy of art is reminiscent of the Native Americans: "Ramon works with the idea of keeping a legacy alive, of storytelling and of sharing. He doesn't hold back because the idea is to move forward." And this guy is going full throttle.

Willing to Change

Like the ocean tides, Camarillo's demeanor remains one of never-ending energy, but he's a bit less predictable, more open to exploration and discovery. He talks about ceramic art with some irony, heaping on respect, while treating it like an unsanctioned science experiment going on behind the high-school gym. It is a healthy relationship—one that is filled with both disappointments and breakthrough surprises.

Given his stature (he's around 5½ feet tall), it's hard to match the man to the medium. That's because he throws big. Very big. In-your-face big. Pots big enough to hold their own in the lobbies of banks and five-star hotels, or in the grand foyers of the homes of the rich and famous, which includes baseball player Mark McGwire.

Throwing forms upside down allows Camarillo to leave the base (which will be the top) a bit thick for support while throwing the bottom; the top is thinned when he finishes the form in the upright position.

His pots are also a combination, a marriage of three cultures: Japanese, Hawaiian and, well, East Coast. They combine the traditional methods that he learned in Hawaii with new tools and techniques—for example, his sparse toolbox contains the standard, tried-and-true wooden ribs, along with an old credit card. He hand burnishes some pots, while others are sprayed with Crystal Clear enamel. He's not a rigid by-the-book kind of guy. On a whim, he will try a new technique suggested at a late-night raku session, or will unabashedly cull ideas from the pots of Otto Heino or Bernard Leach. "They have a lot to show us," he says.

Roots

Camarillo was born in Honolulu and raised in Pearl City on the island of Oahu. His artistic career serendipitously began with a movie. "In high school during lunch, we used to watch surfing movies," he recalls. "There were throwing wheels on the side—lined up against the wall. I was curious, you know, like any kid." Under the tutelage of art teacher Carol Kuahara, he was soon up to his elbows in clay: "I still use some of the techniques she taught me."

But pottery-making was not a full-time endeavor, just a hobby back then. While selling his claywork piecemeal, Camarillo earned an undergraduate degree in marketing and a master's degree in business administration from Hawaii Pacific University.

"I graduated in 1994," he says, "but I couldn't find work in Hawaii. I ended up captaining boats out of Waikiki on dinner cruises. Finally, I decided to come to the D.C. area to find work. My sister lived here, and she sent me a copy of The Washington Post classified ads....It was as thick as a book."

Of his classmates, "I think I've gone the farthest out there," he adds and laughs at the double entendre. "Some of the guys left the islands for California. I went across the country. Hawaiians think it's cold and snowing all the time in the East. It was a shock at first, but I tell them this is a good place to be."

Settling in Vienna, Virginia, Camarillo took a suit-and-tie job at a bank but quickly realized he would rather get his hands dirty with clay than with photocopy ink. Whether he is potting, teaching or performing (he also sings and plays the ukulele in a four-man group called The Aloha

PHOTOS: FRANK RUGGLES, ADAM TASENAN

The first step in making a trash can into a raku kiln is to cut an opening near the bottom for the burner port.

The galvanized trash can and lid are both lined with refractory fiber blanket.

A hole is cut through the lid and the fiber for a chimney.

"Waikiki," 25 inches in height, wheel-thrown stoneware, with glazes, raku fired.

Boys, a traditional/contemporary Hawaiian band), art is his life now.

His job as "resident artist" at two northern Virginia arts centers—Creative Clay Studios and the Lee Art Center—allows him a dialogue with all sorts of clay lovers, from the high-school beginner to the well-seasoned professional to the serious patron. "Ramon's workshops always sell out," says George Brown, director of the 4500-square-foot Creative Clay Studios. "Why are they so popular? Because Ramon's personality is outgoing and magnetic. He has an uncanny ability to creatively share information."

Always Something New

During a morning session at Creative Clay, Camarillo referred to a computer-generated picture of a large, stylized Grecian urn with a tiny, almost nonfunctional foot that he had seen in a magazine. After looking at the picture rightside up, he realized if he turned it upside down, he might be able to reproduce the shape.

He began by slapping a 25-pound mound of clay—instead of raku clay, he uses a commercial stoneware (Soldate 60) right out of the bag—onto an electric wheel. As he threw, the clay moved up, but not out, to form a thick cylinder, about 30 inches tall, with a diameter of about 8 inches. It was so tall his armpit practically became a tool to compress the rim. Initially, there was a lot of water splashing about, but once he had thrown and collared the clay to its full height, he did not add any more, which accounts in part for the pot's integrity and ability to soar.

Watching him work was mesmerizing; watching him watching his work was equally riveting. His gaze followed the graceful curves—hulas, he calls them—of the clay as it began to take shape, checking, rechecking, contemplating, planning ahead, envisioning the design.

His intention was to shape the pot so that the top would become the flat bottom of the vessel. The base would eventually serve as the top of the finished vase, so he left 2 inches of wet clay to be thrown the next day. Once the top-cum-bottom half was dry enough, he flipped the pot over, reattached it to the wheel head, and started throwing the rim of his Grecian urn. The final piece stands about 30 inches high.

Burnishing and Glazing

To burnish his pots, he uses a rather unconventional method. First, while the leather-hard pot rotates on the wheel, he sprays it with water. Then, holding a metal rib in his right hand and a blowtorch in his left, he compresses the surface with the rib as the heat from the torch evaporates the moisture. Final touches are given with, of all things, a shot glass that he rolls over the surface. In Camarillo's studio, anything is up for grabs as a potential tool!

Texture is added not by imprinting, carving or adding clay, but by layering glazes on bisqueware so that effervescent drips bubble and run across the smooth surfaces. Working with just five slips and glazes, he puts "them on thick. I put them on thin. I overlap them. I try them with underglazes. You could spend an entire lifetime with just five glazes and get fantastic results."

Firing

In Hawaii, Camarillo fired his work using the open-pit method. In Virginia, he uses a trash can kiln, which is economical, both in construction cost and in controlling heat loss. "I can get ten firings from one can of propane," he notes.

"Lithium Moon," 16 inches in height, wheel-thrown stoneware, with glazes, raku fired, by Ramon Camarillo.

If one is available, he uses an electric kiln to preheat the glazed pots. This is important, because he wants to make sure they are as dry as the desert; no moisture means better results and less cracking.

Meanwhile, the raku kiln is also heating up. He brings the pots up to temperature "real fast." After five minutes on the kiln shelf, "I turn up the heat. I watch to see the flame coming out of the top of the kiln—out through a little chimney hole. I also listen; the sound that the gas flame makes can tell me when the pot is ready."

Once the glazes have melted, the pot is taken out with a pair of tongs and quickly transferred to a nearby trash can filled with shredded newspaper. The lid is immediately clamped on. He's proud of the fact that his firings release very little smoke.

From soup to nuts, a single pot takes from 30 to 45 minutes to fire, which allows Camarillo to complete about nine or ten firings in a day. "Fast and furious would describe the process," he says.

Given the size of Camarillo's works, it is amazing they are wheel thrown in one piece; most artists would construct such large vessels by joining two or three thrown sections or by handbuilding. More amazing is that they are raku fired at cone 06. His breakage? Zero percent. "After all these years," he says, "I know what not to do. The only breakage I get is when I drop one carrying it home," he says ruefully, describing a particularly painful incident when he lost a beloved pit-fired pot. "It was sad—I felt like crying—but what are you gonna do? It is a pot. I can make it again."

Recipes

White Crackle Glaze

Cone 06

Ferro Frit 3134	85 %
Kaolin	15
	100 %

Luster Duster

Cone 06

Ferro Frit 3134	67 %
Nepheline Syenite	33
	100 %
Add: Copper Carbonate	10 %
Tin Oxide	5 %
Bentonite	2 %

Black Copper Luster

Cone 06

Ferro Frit 3134	75 %
Nepheline Syenite	25
	100 %
Add: Black Copper Oxide	15 %
Black Iron Oxide	15 %

Lithium Slip

Cone 06

Lithium Carbonate	29 %
Kaolin	14
Silica	57
	100 %
Add: Copper Carbonate	4 %
Bentonite	3 %

Red Green Glaze

Cone 06

Ferro Frit 3134	50 %
Custer Feldspar	50
	100 %
Add: Copper Carbonate	3 %

Michael Lancaster
Reinventing Oneself

by Hollis Walker

Connectivity, 8½ inches in height, thrown, altered and assembled clay, raku fired with ferric chloride fuming.

Michael Lancaster put his career as a ceramist in a holding pattern while that of his wife, Barbara Harnack, soared for the last two decades (see "Barbara Harnack: Archtypal Tribe," Raku, Pit & Barrel). After leaving their successful pottery in Malden Bridge in upstate New York and moving to New Mexico in 1987, Lancaster said all he could really do was to be Barbara's assistant.

He never stopped making ceramics; it's just that, much of the time, he was wedging and pugging clay, throwing raw vessels that his wife would later ornament, packing and shipping her creations to galleries across the country, and helping represent her at major craft fairs. And then there were the large adobe house, guest house and studio the couple built on their Cerrillos, New Mexico, property south of Santa Fe: Lancaster was chief builder and contractor on the job for twelve years, although Harnack did her share of adobe-building as well.

The house was completed in 2005, and perhaps Lancaster would soon have begun to refocus on his own work, but life has a way of changing swiftly. In the fall, Harnack and their 17-year-old daughter were in a serious car accident. Though neither suffered major injuries, Harnack sustained serious soft-tissue damage and simply couldn't keep up her previous pace.

At the craft shows in 2006, the couple purposely took fewer wholesale orders for Harnack's work to ease her work load. "But then we couldn't pay the bills," Lancaster recalled. "She needed to slow down, and I needed to come up with something of my own again," Lancaster said. "I saw that as an opening."

He recalled the thought processes that accompanied the economic demand: "So here I am, at age 51, I'm pretty proficient in clay—I can make what I set out to do." But he didn't want to make his old favorites—soda-fired vessels and Asian-influenced raku tea bowls in a soft organic palette—or other common ceramic forms at which many other people are talented.

Unbeknownst to Lancaster, the

Push Pull, 19 inches in width, thrown, altered and assembled clay, raku fired with ferric chloride fuming.

influences of building their home, both the engineering and the construction, had been germinating in the back of his mind. He picked up an idea he had explored many years ago, and began developing it into the Habitat Series. These nonfunctional vessels, essentially cylindrical forms with added feet and funnel-shaped tops, reference ancient forms of residence ranging from lighthouses to Navajo hogans.

One evening, he found himself thinking of his former mentor with whom he had apprenticed, stoneware potter Douglas Conkling. Conkling had Lancaster make thousands of pieces, cut them into parts, and reassemble them into different functional forms. "So at one point when I was really struggling, I grabbed this bow saw and cut one piece [of leather-hard greenware] in half," Lancaster recalled. He reconfigured the two halves at a 45 degree angle to each other and reconnected them. He liked the results and kept experimenting.

He discarded some of the early split pieces. "They were really bad; they didn't work visually," he ex-

Relic, 19 inches in width, thrown, altered and assembled clay, raku fired with ferric chloride fuming.

plained. But as he continued to bisect his ceramic forms, Lancaster intuitively recognized the echoed influences of the chopping and cutting of ceramists Paul Soldner and Peter Voulkos: "These are the people who really influenced me."

Soon Lancaster became inspired by the scrap metal pieces leftover from the nineteenth and early twentieth-century mining activities near his home. Friends and neighbors often discover the old rusty fragments of machinery on their property or on hikes, and place them in their yards as lawn ornaments. Lancaster saw them as mysterious sculptures. "You don't know what it did... but you know it did something," he said.

Those heavy hunks of steel led to Split Cans, Split Canisters and Ancient Industries, ceramic sculptures based on the forms he saw; things that looked like old cam shafts, or automotive mufflers or giant gears. Suddenly, he realized, his works had ceased being vessels at all and had become purely sculptural. The new works weren't replicas of the forms but were inspired by them.

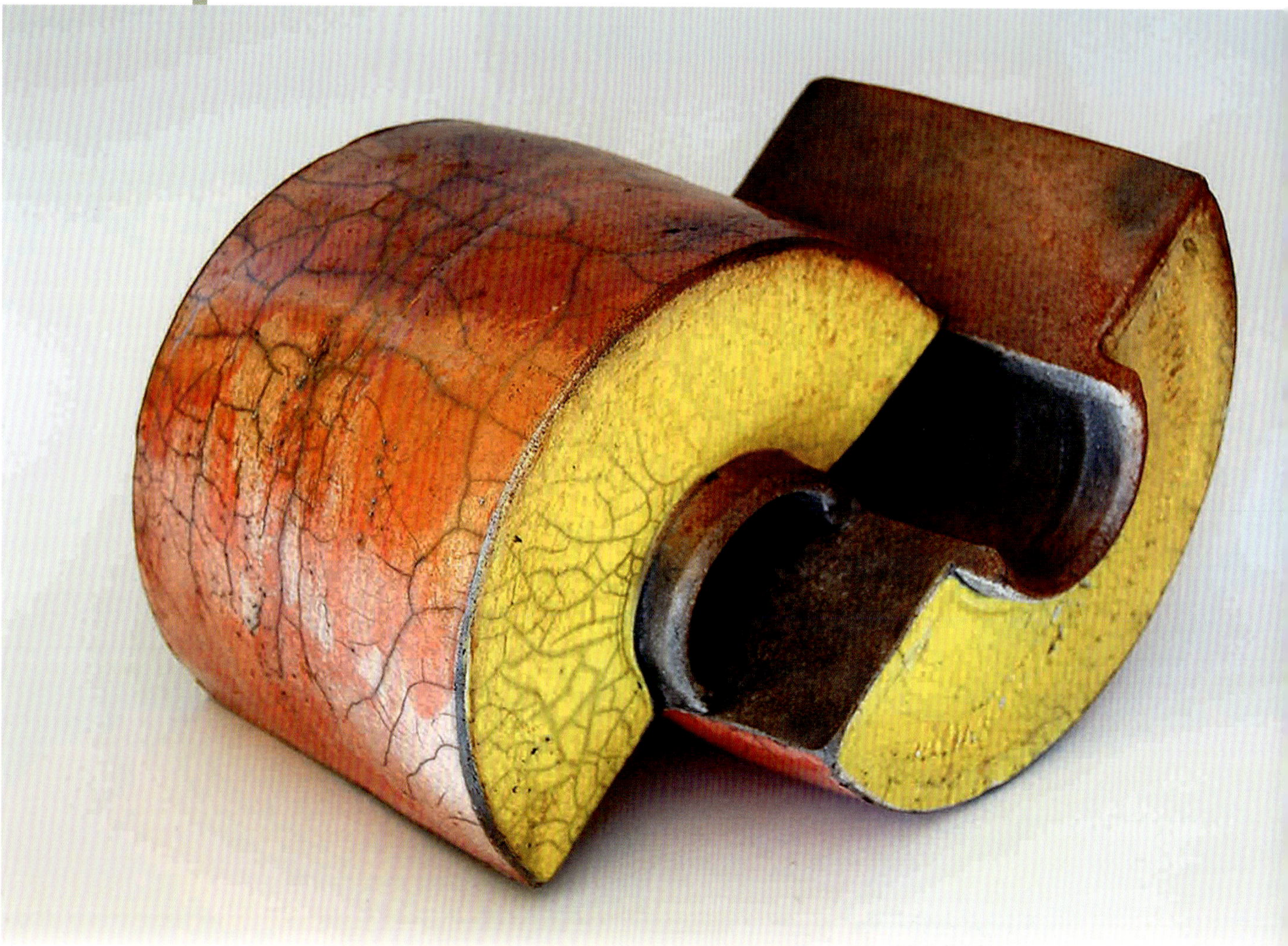

Split Canister, 7½ inches in width, thrown, altered and assembled clay, raku fired with ferric chloride fuming.

"I don't want to make something that's a reproduction of something else; I want the viewer to get engaged the same way I do with the stuff in the salvage yards," he explained. "Wouldn't it be ideal if you could get up every day, and look at something and question its history? If things like the yard scraps aren't always beautiful, they're always intriguing."

Lancaster throws his forms on the wheel, then slices them into segments and rejoins them, or attaches other pieces to them. He leaves the cut surfaces rough. "I want people who are into ceramics to know it's a wheel-thrown object that retains its clay qualities," he said. The strong New Mexico sun speeds the making process; he often places the pieces outdoors to dry them faster so he can cut them up. Once they are reconstructed, bisqued and glazed, he preheats the works to 300°C (570°F) before putting them in the large outdoor raku kiln he and his wife share. Of late, he's been using brilliant glazes—lavenders, blues, greens and yellows—and iron-fuming the pieces to make them look old. On top of the glazes, the iron fuming creates a luminous, age-polished patina. One of his latest favorites is Push-Pull, a long cylinder that appears to have a smaller, longer cylinder pushed through it at an angle.

Lancaster's new works have more in common with the sculptures of Mark di Suvero, whose work he admired during several visits to Storm King Art Center in Mountainville, New York, than with contemporary ceramic sculpture. Juxtaposing his rounded forms with the geometric planes of their sliced surfaces, the new works appear authoritative and weighty, a sensibility that is balanced by their playful colors. Though the impulse to bisect his forms shares an affinity with the practices of Soldner and Voulkos, Lancaster's pieces lack their organic messiness, instead favoring a sleek, contemporary aesthetic. Lancaster comes by his creative passions naturally. His mother, Betty Warren, was a well-known portraitist, and her father, Alonzo "Jack" Warren, was the co-creator of the cartoon character Pecos Bill. Lancaster's great-grandfather on his paternal side was Charles Ringling, of Ringling Brothers fame. In addition to making ceramic art, Lancaster is an abstract painter, selling his paintings through galleries separately from his clay works.

But the return to creating in clay has been exhilarating and profitable as well. "I'm having a blast. I have all the passion I had in that first year I made ceramics, 1976, but with a sense of confidence," he said. "I draw all my new pieces on newsprint—big or bigger than life-sized—and then execute what I have drawn. Sometimes I wake up in the morning and imagine an invention, a new machine or an old industrial relic. Usually I try to make them within a day or two. They don't always work, but each is an exercise in growth. Lately, they are selling as fast as they arrive in galleries, so I don't feel pressured—just the thrill of keeping up."

In the meantime, Harnack's health has much improved, and the two are both working steadily in their favorite medium. Lancaster has gained a new perspective on the revival of his career, one he feels is worth sharing.

"Sometimes, when we're 25 or 35 years into our career, we have to reinvent ourselves," he said. When the couple needed money, he realized he could have gone into construction and safely met the bills. Instead, he took the bigger risk. "I decided to stay home and do what I love to do," he said. "And it worked."

Raku Safely

by Michael Lancaster

I look back to 1977, the first year I ever fired American raku. I think, secretly, "It's a miracle any of us have survived this long!" We were fuming with petrochemicals, smoking with sawdust, newspaper, pine needles, straw and even unknown combustible garbage. Working with clay can, for a potter or a sculptor, be a romantically "macho" sort of thing, especially when it comes to dancing with fire.

In this age of online videos, the ceramist can view many different approaches to firing and safety. I have seen some that are downright scary. For example, wearing paper dust masks for lung protection, wearing nylon jackets, synthetic jogging outfits, shorts, sandals and rarely do I see a fire extinguisher in the background. From experience I have found that clothes can ignite, hair gets burned off, hands burn, and lungs and mucous membranes can be damaged —even permanently! I have come up with a few simple items that can help make raku more safe.

First, to protect eyes, face, mucous membranes and lungs, I recommend Color Code Olive/Magenta P100 Defender respirator cartridges from North, or another company using the same codes, in a welder's clear plastic face and respirator shield. The cartridges will arrest the fumes and smoke from welding and will help for most of the smoke from Raku. The plastic lenses generally are not fireproof, however will shield from some flame up and will keep the smoke and fumes from your eyes. At a minimum, care should be taken to be sure all clothes are cotton. We also wear split-hide welders jackets. Some of my colleagues wear full-hide aprons that go all the way to the shins. This can be helpful for extended moments reaching with mits into the kiln.

We also wear different gloves for different tasks. I wear Kevlar gloves for general work and they last about three firings (that's about five to six firings each time from a 30-cubic-foot kiln). I use a higher-temperature Kevlar combination fiber mitt for hand lifting works that are too big for the tongs. Boots with more leather and less plastic are helpful as I have had boots burst into flame.

When fuming, it is an absolute must to wear good lung protection. It is extremely important that anyone within a 30-foot circumference also be protected. Fuming ingredients such as ferric chloride, ammonium chloride, copper sulfate, etc., make caustic gases which can burn sinuses, throat, esophagus and lung tissue and can cause permanent damage. Finally, always keep a fire extinguisher on hand. It could save a life. Raku should be fun, passionate, and, with a calm approach, safe.

Standing next to the 30-cubic-foot raku kiln, Lancaster demonstrates full respiratory protection. The blue mits are for removing large works from the kiln. In his hands, he is holding Kevlar gloves, which are used for turning the hot pieces over in the reduction barrels.

While this is a dramatic image, Lancaster inhaled a large quantity of smoke during this firing, resulting in bronchitis. This was his last firing without respiratory and eye protection.

The Nichrome Wire Solution

by Cory Pinassi

"Urn," 11½ inches in height, raku clay, with lithium glaze.

Whether by circumstance or design, we live our lives in boxes of varying size, shape and texture. In the English language, we speak of parameters, limits, being "boxed in" or "blowing the lid off." All these references are metaphors for the constraints of various aspects of our lives. Our homes, vehicles, theaters, gardens are defined by perimeters and parameters.

We are even confined by the geography and class into which we are born. Some boundaries can be extended or removed, whereas some walls, such as the century in which we are born, our body type and color, remain immutable. Whether or not we choose or are able to alter these definitions of our life, containers define who we are as people.

My clay containers are handbuilt. Early training in cloth and clothing design led me to approach clay as a surface to be cut and assembled. After rolling a slab, I then cut and assemble the parts and continue manipulating them until achieving the resultant form.

My preferred firing technique is raku, to which I was introduced in 1969 while attending a large potluck Thanksgiving dinner in Los Angeles. One guest came with a stack of bisqued bowls, two buckets of glaze and a handful of brushes. Each of us was given a bowl and instructed to decorate it, with the assurance that it would be fired and returned. Having taken a couple of pottery classes in college, I wondered how he would track us down after the three days it took to do a firing.

As the rest of us were enjoying our dessert, the pottery-guest built a huge fire in our hosts' oversized fireplace, into which he randomly placed our decorated bowls. At one point, there was an urgent shout of "open the door!" We turned to see a man flying through the front room with a glowing-hot bowl at the end of a pair of fireplace tongs. In amazement, we followed him outside and watched as he placed the bowl in a pile of leaves in the corner of the patio, quickly covering it with an old washtub. And so it went, repeatedly, until we each had our black and red or yellow bowl (lead glazes were still common at that time). We were shocked, amazed and delighted.

Clay reentered my life in a 1982 evening class at Sheridan College of Craft and Design in Mississauga, Ontario, Canada. It all seemed vaguely familiar—wedging and throwing—until the teacher said we would be doing a raku firing on the last night of classes. Nothing twigged until the teacher pulled a red-hot piece from the kiln and placed it into a lidded trash can. Bingo! The lights went on, and the thrill was back in clay!

With family responsibilities and recovery from auto accident injuries in the forefront, clay sat in the background until I took continuing-education classes at Sheridan College in Oakville, Ontario. Once again, the last class was a raku firing, and afterwards, my affair with clay was again in full flame. At the suggestion of my teacher, Wendy Vervoort-Kline, I joined the Mississauga Potters Guild, which has a facility at which I could work.

The guild tries to organize one, and sometimes two, raku firings per year. By participating in the group firings, I slowly began to understand the results of various glazes and firing techniques. It was also helpful to observe the actions of others and their results.

Many years ago, a few of us began firing together more often. Eventually, we organized a one-day gallery-style exhibition at the local community arts center (Springbank Visual Arts Centre in Mississauga). While we were trying to come up with a name for ourselves, my daughter suggested Fireballs—and it stuck. Eventually, we built our own kilns, but we continue to meet, exchange information, encourage each other and plan exhibitions together.

A successful raku firing begins long before the kiln is turned on. It is always frustrating to go through the whole process of conceptualization, construction, drying, bisquing and glazing only to lose a piece in the raku firing. To facilitate its removal from the hot kiln, the form usually needs to have an opening into which tongs can be inserted, handles under which tongs can be slipped or a diameter narrow enough to squeeze around. Any of these methods can prove to be stressful on an object that is glaze-melting hot.

As my containers became larger and more streamlined, it became harder to remove them from the kiln to the postfiring reduction chamber without serious damage or loss. Also, I wanted to leave their lids in place during the firing process to ensure a consistent result over the whole surface.

I tried the method of wearing fire-resistant gloves to transfer pieces from the kiln to the reduc-

"Memory Box," 14½ inches high, by Cory Pinassi.

tion chamber by hand, but the piece was usually too close to my body and the gloves quickly burned through. Ruining gloves that fast was way too expensive.

Then one day, while using some Nichrome wire to replace some buttons securing fiber to the lid of my kiln, I was struck by how flexible and long-lasting this wire was. So I thought, "Why not just cage the piece with wire, tie it at the top and add a loop to insert the ends of the tongs for touchless removal from the kiln?" It worked like a charm.

Nichrome wire comes in a variety of gauges and can be purchased from most pottery suppliers. The thickness of the wire will be dictated by the weight of the piece being fired, and it is reusable. While it is true that the wire becomes more brittle with each firing, I figure that it is safe to use until it becomes too stiff to bend with my fingers, or it snaps when I bend it. At that point, it's time to make a new cage.

This Nichrome-wire solution has allowed me to raku large, lidded containers with a greatly reduced loss rate. I now use it for any piece that cannot be easily and safely removed from the kiln with tongs alone.

Fitting a lidded piece with a Nichrome wire cage facilitates removal from the kiln for postfiring reduction.

Neoclassical Raku

by Robert Hasselle

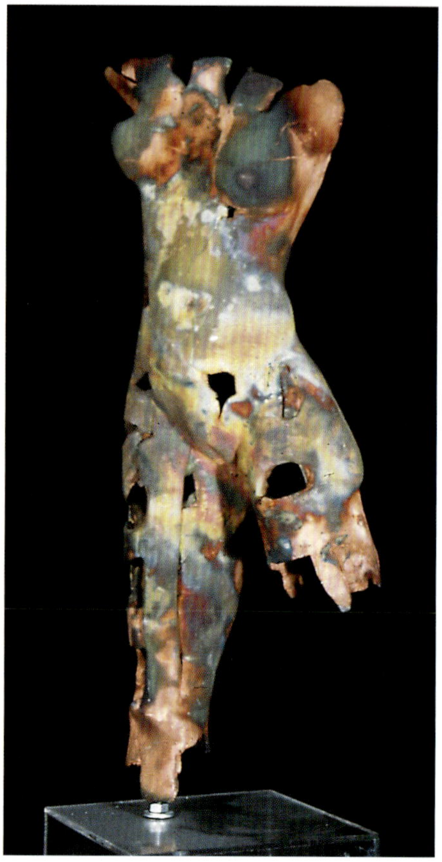

"Reconstructed Fantasy," 30 inches tall, with Hasselle Copper Matt.

After years of production pottery, I found myself needing to get back to my sculptural roots. None of the glaze surfaces I had been working with seemed well adapted to sculpture, so a change was in order. When I encountered the copper matt raku process, it all began to come together.

Copper matt raku produces a surface finish resembling bronze, somewhat metallic in appearance, but more lively. All the tedious processes and finishing work of bronze could be done away with, and yet the surfaces would be aesthetically appealing.

Needless to say, there is a lot of groundwork to do before one can get a handle on the copper matt process. I was lucky enough to attend a Robert Piepenburg workshop at the beginning of my efforts, and thus started off on the right foot.

At first I was content to use other people's recipes, and settled for hit-or-miss effects. Some of these may be a little more colorful than mine, but most of them are too high tech for me. My recipe (developed with the help of Calvin Hubbell) does not require a pyrometer or stopwatch, or any kind of exact schedule in order to work. My guess is that its window of opportunity is somewhat larger.

Do not, however, expect perfection! There is no such thing as total control with copper matts. If your success rate is above 70%, you are doing well.

My copper matt recipe includes ball clay, and unlike the frit recipes, works on the principle that high concentrations of copper act as a flux, and hence do not need a fluxing agent. I think this fact opens up a wide variety of possible glaze ingredients, especially the lower temperature materials like talc and Cedar Heights Redart, but I have not had time to experiment with them.

Some artists use red copper oxide in their copper matt recipes, but I prefer black copper. If you have trouble dispersing the copper in the recipe batch, try adding ⅓ Windex to ⅔ water. That should reduce the surface tension and the copper should mix in readily.

My copper matt vessels and sculptures contrasts with another body of work glazed with a bright blue to purple to aqua barium matt colored with copper. It's interesting that a single colorant can produce such opposite, complementary colors.

After two years of experimentation, I have a few suggestions to offer. I recommend keeping a chart of the factors that affect the copper matt finish—size of piece, temperature on removal from the kiln, type of combustible, condition of flame when the reduction chamber is covered, amount of combustible placed on top, time before removal, humidity, etc. The chart should help you identify the desired conditions and promote success.

One of the first choices you will have to make is to decide whether to build a totally portable kiln or find a space with a roof over it. After all, you may find that rainy days produce the best results. I fire in my garage in any weather.

The second problem is, depending on the degree of enclosure, how to get rid of the smoke. An exhaust fan is desirable in tight spaces.

The choice of combustibles for postfiring reduction is your next concern. I have had success with sawdust and colored newsprint, but dry leaves are best for my glaze. The application of this fuel is also important. Some glazes require lots of fuel and flame; others do better just smoldering. Some potters put a lot of fuel on top of the pot; others just roll the pot around in it.

The next decision is when to remove the pots after turning off the kiln. This is the stopwatch period. Fifteen seconds to a minute is the usual time allowed. If one is using a pyrometer, the 1300°F to 1400°F range is used.

With the pyrometer approach, one immediately transfers the pots to the reduction chamber. With the eyeball approach, one uncovers the pots by removing either the entire kiln or the lid, and watches for the right color on the surfaces—for me, this is in the dark red to burning black range.

The next step is the postfiring reduction itself. Here, you will want to monitor the condition of the flame when the pots are covered. This can vary from just smoking to fully engulfed.

I want the leaves underneath to be ignited by the heat of the pot, which, in turn, should ignite the combustibles sprinkled on top. However, the

"Reliquary Vessel," 42 inches in height, wheel-thrown and handbuilt, with barium matt glaze.

added leaves should not burn so fast that they are entirely consumed before I can cover the container.

Treatment of the pot after reduction varies considerably. Some potters do a "burping" procedure after about five minutes, which means reopening the reduction chamber to allow air to enter. Others remove the pot and quench it in water to "freeze" the color. I believe in simply leaving the pot in the postfiring reduction can for 30 minutes.

To re-enter the world of sculpture after 25 years is to find oneself in a different universe. My earlier work, tongue-in-cheek satires of our national pride, seems totally inappropriate in the current dispirited era.

Currently, my central theme is fragmentation. I see a fragmented society, a fragmented belief system. I think the main problems behind the splintering of America are multiculturalism and the ever-expanding rift between the rich and the poor. Another more subtle destabilizing factor is the dethroning of science in the postnuclear age, giving rise to various forms of New Age quackery. In a sociopolitical sense, the fragmentation of my figures refers to the fragmentation of America. In a more personal sense, I refer to these figures as my male menopausal sculpture series.

In this sense, I am de-mythologizing the role of the artist, and confessing to the same frailties as the rest of humanity. This is a call for a little more honesty and less posturing on the part of artists themselves.

The fragmented torsos, in a way, though, tend to contradict the intact neoclassical vessels. They stand next to each other somewhat in the relationship of "disease" and "cure."

"Neoclassical Amphora," 22 inches in height, with copper matt finish, reduced in leaves, by Robert Hasselle.

The cure, however, is worse than the disease.

Artists with historical perspective are aware that neoclassical imagery is associated with repressive regimes. Hitler's architect, Albert Speer, was an ardent neoclassicist. My fondest hope is that the "Balkanization of America," as Charles Krauthammer has labeled it, will be ended by a renewed consensus of values. If not, it will most likely be ended by force, repression and externally imposed canonical value systems. Nature abhors a vacuum.

Recipes

Hasselle Copper Matt

Black Copper Oxide	80 %
Ball Clay	20
	100 %
Add: Bentonite	3 %

Sunday Copper Matt

Revised

Copper Carbonate	90 %
Ferro Frit 3110	10
	100 %
Add: Bentonite	2 %

Piepenburg Red Bronze

Black Copper	2.4%
Tin Oxide	2.8
Gerstley Borate	47.4
Ferro Frit 3134	47.4
	100.0%

Piepenburg Patina

Copper Carbonate	1 cup
Bone Ash	3
Gerstley Borate	4
Nepheline Syenite	2
	10 cups

Post Pac Man

Copper Carbonate	1 cup
Red Iron	⅛
Bone Ash	1
Gerstley Borate	5
Nepheline Syenite	1
	8 cups

Large Raku Platters

by Tom Radca

Tom Radca uses slips and stains on his large thrown platters. An alternative postfiring reduction technique decreases losses from thermal shock on platters up to 36 inches in diameter.

I've been making pottery for more than twenty years, specializing in large platters and murals composed of hand-cut tiles. My first venture into making large platters ended in disaster. Out of more than seventy large platters, only eight survived. Through experimentation (and determination), I was able to resolve all the technical issues dealing with throwing, manipulating, drying and firing these large forms.

As for glazing, I wanted to get back to the kind of glazes I used at Kansas State University working under Angelo Garzio, but I just didn't want to go up to high-fire stoneware temperatures. So I developed the raku glazing techniques that I now use on my platters. I found a cone 04 glaze that I thought would be good for a raku-type firing, and sprayed it on a vessel, applied a stain of iron and copper, then fired it in an old electric kiln that I had converted to a raku kiln. I've also discovered that the cone 9 white glaze I used in school could be used at the much lower raku temperatures. The under-fired glaze had a cratered lava look and by using iron and copper stains, I was able to achieve surfaces reminiscent of landscapes from my many travels abroad.

Firing large platters using a raku firing method would be extremely challenging if I used a conventional firing regimen, that is, fire to temperature, then remove the work and immerse it in a post-firing reduction receptacle. With platters weighing more than 30 pounds, this method would not only be dangerous, but it would most likely increase the breakage rate. My solution was to place my work in a kiln, fire it up to cone 09, at which point, I shut off the gas and stoke the firebox full with oak. After a good fire has started in the kiln, I close up the flue hole in the top, brick up the burner port and then walk away; no lifting, no hassling with heavy red-hot work, and no breathing smoke. Unloading the kiln the next day nearly always reveals success.

Throwing Large Platters

Use 36 pounds of fairly stiff clay for these large platters (figure 1). To reduce stress on your back and arms, center only 12 pounds at a time. Center and flatten each 12-pound lump before the next lump is added (figure 2). Standing will improve your leverage.

Open up the centered clay and check the thickness of the bottom (figure 3). It should be about ¾ inch thick. Start the platter as a funnel shape before pushing the sides down into a platter form (figure 4).

Trim excess clay from the base of the platter, then remove the bat and platter from the wheel and hold chest high (figure 5). I discovered that it's next to impossible to flip a freshly thrown large platter upside down onto a drape mold. Intentionally collapsing the rim makes it possible, but you have to bounce the bat to get the clay to collapse (figure 6). With the rim completely collapsed, get your shoulder under the rim of the bat to prepare to flip the platter over (figure 7).

If the rim had not been collapsed, it would have folded under when flipped over. Once it's on the form, remove the bat (figure 8). To reduce cracking problems, trim the excess clay from the base to get a more even thickness between the lip and the foot (figure 9).

Alter the form using a rolling pin (figure 10). Since the clay is still very wet at this point, use plastic as a barrier so it doesn't stick to the pin. Roll clay from the rim toward the base so that the foot becomes a dome (figure 11). Using your thumb, press in on the center of the dome to create a foot (figure 12). This will allow the platter to sit flat.

Once all the forming is complete, cover the lip of the platter with plastic so it will dry slowly (figure 13). The plastic should stick because the clay is still wet, but it will release as the clay dries. Use a bent nail to make holes for a hanger (figure 14). Depending on the time of year, these platters can take anywhere from a few days to a couple of weeks to dry—don't rush it.

Recipe

Kansas State White Slip

Cone 9

Dolomite	10 %
Custer Feldspar	14
Ball clay	29
EPK Kaolin	38
Silica	9
	100 %

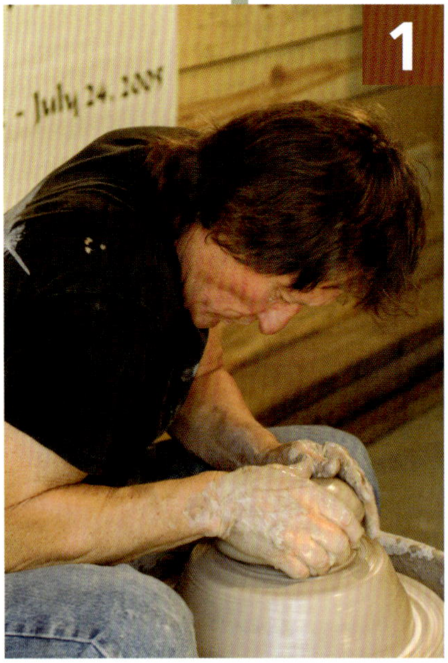

TIP

Using a steel yardstick on the inside brings the wall down quickly and evenly.

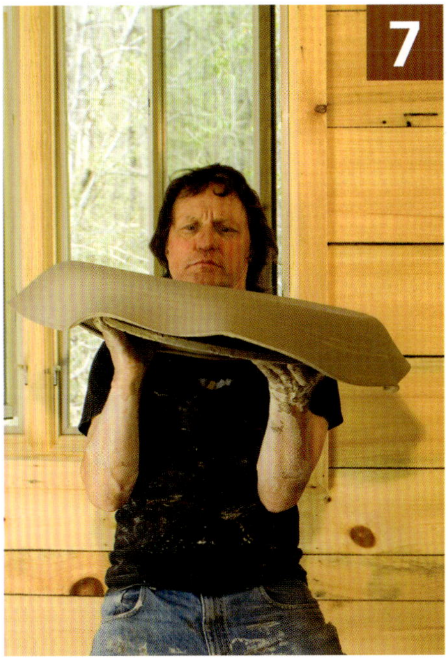
7

8

The form that I flip my platters onto is made from the hood of a Vent-A-Kiln, which I ordered without a motor or hardware. You can also use a child's saucer sled. Cover the form with enough plastic to extend 6 inches beyond the platter you're making.

9

10

11

WARNING

Do not use this firing method with an electric kiln.

Place the glazed platter in a kiln. Once it reaches cone 09, remove the burner, stoke the kiln with ten pieces of dry split oak (figure 15), then plug up the burner port. Once the fire is going strong (figure 16), cover the top hole and walk away. The kiln is now a postfiring chamber. When using the kiln for both firing and postfiring reduction, large work remains safely in place until it is cool enough to handle (figure 17).

The indentations on this platter were made by pushing a board up under the plastic while the platter was still on the form after the foot was formed. The pits came from placing rock salt on the surface before flipping the platter onto the form. The rock salt dissolves leaving the pit. The colors on this platter were achieved with a stain consisting of 50% iron oxide and 50% copper oxide.

Cone 9 Kansas State White Slip applied thick with iron and copper stains applied. The thick lava effect results from underfiring the cone 9 glaze.